MAKING SENSE OF STATISTICAL STUDIES

Teacher's Module

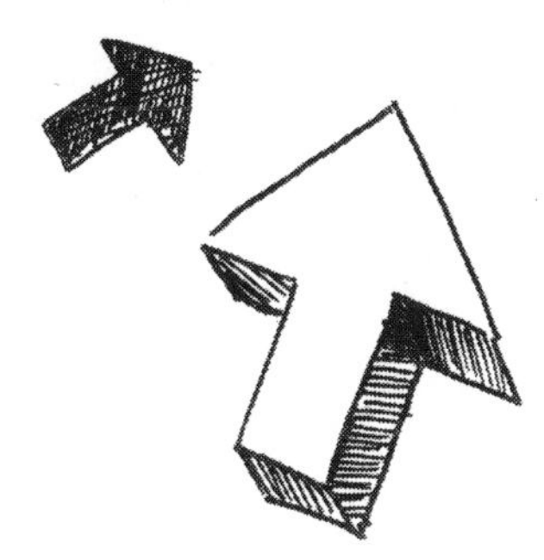

Roxy Peck
Cal Poly, San Luis Obispo

Daren Starnes
The Lawrenceville School

with
Henry Kranendonk
Milwaukee Public Schools Mathematics Curriculum Specialist

June Morita
University of Washington

Production Team

Martha Aliaga
Advisor

Nicholas Horton
Production Administrator

Henry Kranendonk
Contributor

Jerry Moreno
Advisor

June Morita
Contributor

Megan Murphy
Illustrator

Rebecca Nichols
Illustrator

Roxy Peck
Advisor

Valerie Snider
*Designer and
Editorial Production Manager*

Daren Starnes
Executive Editor

©2009 by American Statistical Association
Alexandria, VA 22314-1943

Printed in the United States of America
10 9 8 7 6 5 4 3 2 1

978-0-9791747-6-6

Library of Congress Cataloging-in-Publication Data

Peck, Roxy.
    Making sense of statistical studies / Roxy Peck, Daren Starnes ; with Henry Kranendonk, June Morita.
      p. cm.
  ISBN 978-0-9791747-6-6 (alk. paper)
  1. Mathematical statistics--Study and teaching (Secondary)--United States. I. Starnes, Daren S. II. Title.
  QA276.18.P43 2009
  519.5--dc22

                      2009009395

I AM HONORED TO WRITE THE FOREWORD FOR *Making Sense of Statistical Studies*, a capstone experience for high-school students. As statistics is increasingly recognized as a necessary component of the high-school mathematics curriculum and other high-school curricula, there is an urgent need for materials such as *MSSS*.

In its *Principles and Standards for School Mathematics* (*PSSM*, 2000), the National Council of Teachers of Mathematics (NCTM) articulates a vision for mathematics education that includes data analysis and probability as one of five major content strands. To support and further elaborate on the Data Analysis and Probability standard, the American Statistical Association (ASA) produced the report *Guidelines for Assessment and Instruction in Statistics Education (GAISE): A Pre-K–12 Curriculum Framework*. The goals of the GAISE Report include the following:

Presenting the statistics curriculum for grades Pre-K–12 as a cohesive and coherent curriculum strand

Promoting and developing statistical literacy

Providing links to the NCTM Standards

Articulating the differences between mathematical and statistical thinking—in particular, the importance of context and variability with statistical thinking

Clarifying the role of probability in statistics

Illustrating concepts associated with the data analysis process

The GAISE Report was endorsed by the ASA in 2005 and, with funding provided by the ASA/NCTM Joint Committee on Curriculum in Statistics and Probability, appeared in print in 2007.

The GAISE Framework was developed to help identify the essential topics and concepts in statistics and probability in grades pre-K–12 for all students as they progress from pre-kindergarden to graduation from high school. The Framework has become an instrumental document in defining the role of statistics within the school mathematics curriculum in national documents such as *College Board Standards for College Success: Mathematics and Statistics* (2006) and NCTM's forthcoming report *Focus in High School Mathematics: Reasoning and Sense-Making* (2009). The GAISE Report provides guidance for curriculum directors, faculty of teacher preparation colleges, and pre-K–12 teachers involved with statistics education. Finally, GAISE has had an effect on state standards and writers of assessment items. For example, the data analysis

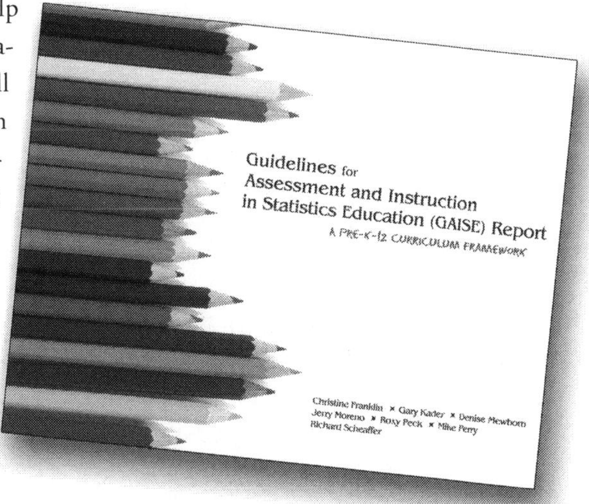

and probability strand is a major component of the new Georgia Mathematics Performance Standards.

The GAISE Report submits that every high-school graduate should be able to use sound statistical reasoning to intelligently cope with the requirements of citizenship, employment, and family and to be prepared for a healthy and productive life. It also holds that statistics is a must-have competency for high-school graduates to thrive in this modern world of mass information. Taken together, *PSSM* and the GAISE Report describe statistical problem solving as an investigative process that involves the following four components:

> Formulate a question (or questions) that can be addressed with data
>
> Design and employ a plan for collecting data
>
> Analyze and summarize the data
>
> Interpret the results from the analysis and answer the question on the basis of the data

The GAISE Report outlines a conceptual structure for statistics education in a two-dimensional framework model with one dimension defined by the statistical problem solving process, plus the nature of variability. The second dimension is comprised of three levels of statistical development (Levels A, B, and C), which students must progress through to develop statistical thinking. Grade ranges for attainment of each level are intentionally unspecified. It is paramount for students to have worthwhile experiences at Levels A and B during their elementary and middle-school years in order to prepare for future development at Level C during the secondary level. A high-school student with no prior experience with statistics will need experiences with concepts from both Levels A and B before moving to Level C.

Making Sense of Statistical Studies is an excellent set of investigations that supports the spirit of the GAISE Framework recommendations and is ideal for a capstone experience of the high-school student who has evolved through Levels A and B and is on to Level C. The investigations in *MSSS* bring the real world to the student and provide students the opportunity to understand the necessity of statistical reasoning and sense-making for everyday life and post-secondary education.

I'm most grateful to the writers of *MSSS* and the ASA/NCTM Joint Committee for developing this valuable resource in support of both the recommendations of GAISE and the importance of statistical reasoning in our high-school curriculum. These dear colleagues share the vision of promoting statistical reasoning for all high-school graduates.

Christine Franklin
Chair of the GAISE Report for Grades Pre-K–12

MAKING SENSE OF STATISTICAL STUDIES (*MSSS*) IS DESIGNED AS A STAND-ALONE experience with the methods of designing and analyzing statistical studies. It is written for an upper middle-school or high-school audience having some background in exploratory data analysis and basic probability.

The *MSSS* student module consists of an introduction and four distinct sections. Each section begins with an overview that contains essential background information for students. The remainder of the section is devoted to guided student investigations. These investigations start with a research question on some topic of interest. Students are then led through a series of questions that help them examine the study design, analyze data, and interpret results. Later investigations ask students to design, carry out, and analyze results from their own studies. A description of each section follows.

The **Introduction** gives students a snapshot of the statistical problem-solving process. It includes a brief discussion of the primary methods of data production—surveys, experiments, and observational studies—as well as the difference between a sample and a population. Ethical issues involved in data collection are also mentioned here.

Section I: Observational Studies shows students how much can be learned just by watching and recording data. The first two investigations in this section help students review the primary graphical and numerical tools for analyzing data. The remaining investigations incorporate random selection, which allows students to generalize the results of their data analysis to some larger population of interest.

Section II: Surveys begins with two investigations that require students to examine data from surveys and critique the design of surveys that have already been conducted. The questions include a review of some basic ideas of probability. In the final investigation of this section, students are led through the process of administering their own survey.

Section III: Experiments starts with an investigation in which students practice using the terminology of experiments as they review the details of two studies involving dieting and weight loss. In the next investigation, students are guided through the process of designing an experiment to test the effect of listening to music on memory. Once they have collected the data, students must use the data analysis and interpretation skills they developed in Section I to help answer the research question. Students get to design, execute, and analyze results from their own experiments in the final investigation of this section.

Section IV: Drawing Conclusions introduces students to the basic ideas of inference—estimating a population characteristic and testing a claim about a population characteristic. Simulation is used to quantify the sample-to-sample variability that occurs in repeated random sampling. This chance variation is reflected in the margin of error for an estimate and in the decision-making process for evaluating the validity of a claim about some population characteristic.

Acknowledgements

We would like to thank the American Statistical Association (ASA) for its steady support over the past five years as we have worked on this project. The initial funding for *MSSS* came from an ASA Member Initiative. With the publication of the GAISE Report, the ASA took a leadership role in promoting statistical literacy in the pre-K–12 mathematics curriculum. *MSSS* is designed to support the curriculum framework proposed in the GAISE Report.

Our special thanks go to the ASA/NCTM Joint Committee on Curriculum in Statistics and Probability (JC) for its unwavering commitment to making *MSSS* a reality. At a key point in the project, the JC provided additional funding that allowed the lead authors to meet for three uninterrupted days of writing. Under the leadership of Jerry Moreno, the JC also assisted in reviewing drafts of the student and teacher modules to get *MSSS* into production. We offer a heartfelt thank you to our colleagues who served as reviewers: Martha Aliaga, Brad Hartlaub, Dan Lotesto, Deborah Lurie, Jerry Moreno, Tom Short, and Jeff Witmer.

We particularly wish to acknowledge the contribution of Wanda Bussey and her students from Rufus King High School in Milwaukee, Wisconsin, as well as the students from Katedralskolan in Lund, Sweden, for the survey and data provided in Investigation #15.

Two individuals deserve our heartfelt appreciation for their tireless efforts during the final stages of the project. Nicholas Horton managed all aspects of the production process with remarkable professionalism. His close attention to detail and personal touch were essential in resolving many last-minute issues. Last, but certainly not least, Valerie Snider from the ASA deftly coordinated all elements of the design and production process. Her creative flair and passion for the project is apparent in the final publications. For those late nights and long weekends, and for her uncanny ability to meet seemingly impossible deadlines, we owe Val our sincere gratitude.

Roxy Peck
Daren Starnes
Henry Kranendonk
June Morita

Introduction to the Teacher's Module

THE *MAKING SENSE OF STATISTICAL STUDIES* TEACHER'S MODULE INCLUDES SUPPORTING resources to help you use *MSSS* effectively in your classroom. To make the Teacher's Module easier to use, we have included all the pages from the Student Module. Prior to each section and each investigation in the student module, you will find Teacher Notes that provide the following:

A descriptive **Overview** that discusses the big ideas presented in the section/investigation

Any **Prerequisites** for knowledge and skills students will need to complete the section/investigation successfully

A list of **Learning Objectives** that describes what students ought to know and be able to do as a result of completing the section/investigation

Detailed **Teaching Tips** from the authors that include in-depth discussion of key terms and concepts, examples, and specific questions from the student investigations

Source **References**, when applicable

Possible Extensions of the material discussed in the section/investigation in case you want to go further with your students

In addition, we have provided **Suggested Answers to Questions** for each of the 15 investigations.

We have designed *MSSS* to give you considerable flexibility in using the section overviews and investigations with your students. The overviews were written for a student audience. You will have to decide whether to ask students to read them ahead of time in preparation for a class discussion or to have students read some or all of the narrative aloud during class. Likewise, the investigations can be completed entirely during class time or split between class work and homework. For planning purposes, the table below shows our estimates of the time required to complete each investigation.

Section/Investigation	Estimated Time	Section/Investigation	Estimated Time
Introduction & Investigation 1	2 days	Investigation 9 plus data collection	2-3 days
Section I Overview & Investigation 2	3 days	Section III Overview & Investigation 10	2 days
Investigation 3	1-2 days	Investigation 11 plus data collection	2 days
Investigation 4	1-2 days	Investigation 12 plus data collection	2 days
Investigation 5	2 days	Section IV Overview & Investigation 13	2-3 days
Investigation 6 plus data collection	2 days	Investigation 14 plus data collection	2 days
Section II Overview & Investigation 7	2 days	Investigation 15 plus simulation	1-2 days
Investigation 8	1-2 days		

What about assessment? Student responses to the individual questions within an investigation can give you valuable feedback about their developing understanding. We have identified questions in the Teacher Notes as possible assessment tasks. Investigations 6, 9, 11, 12, 14, and 15 require students to complete all four steps in the statistical problem-solving process: defining a research question, collecting data, analyzing the data, and interpreting results. These more comprehensive investigations could be used to formally evaluate students' mastery of the stated learning objectives. Alternatively, you might use one of the suggested Extensions as the basis for a performance assessment. Of course, you could develop and administer more traditional quizzes or tests covering the terminology, methods, and ideas of a given section or investigation.

THE INTRODUCTION IS DESIGNED TO GIVE STUDENTS A GLIMPSE OF THE STATISTICAL problem-solving process. Several specific research questions that students will answer during later investigations are presented in the opening paragraph. We then offer an example that shows how researchers attempted to get an answer to the question: Do most people wash their hands after using the bathroom? Results from a survey and two observational studies are presented. Following the example, we outline the four steps in conducting a statistical study:

Defining a research question

Collecting data

Analyzing the data

Interpreting results

Along the way, we explain how information from a carefully chosen *sample* can be used to make inferences about a larger *population*. In addition, we distinguish the three primary methods of producing data—*observational studies*, *surveys*, and *experiments*.

In Investigation #1: Did You Wash Your Hands?, students answer a series of questions that help them make sense of the hand-washing studies described in the Introduction.

Prerequisites

None

Learning Objectives

As a result of completing the Introduction, students should be able to:

Distinguish between a population and a sample

Briefly describe the three primary methods for producing data

Identify the four steps in conducting a statistical study

Critique information provided in an article or headline

Teaching Tips

Give students practice defining their own research questions. Statistics questions require some variability from individual to individual. For instance, "On average, how far from school do students in this class live?" is a statistics question. Students in the class live different distances from school, so individual-to-individual variation is present. Contrast this with the question, "How far does Tyler live from school?" This is not a statistics question, because there is no individual-to-individual variation present. "What percent of students own an iPod?" is a statistics question, while "Does Kayla own an iPod?" isn't.

For each acceptable research question that a student defines, ask the class which method of data collection—observational study, survey, or experiment—would be optimal.

Then, ask students to give some additional examples of research questions that would be best answered using each of the three methods of data production.

Ask students why researchers don't just get data from every member of the population—a census. Some issues that make a census impractical include the difficulty of contacting some individuals, time constraints, measurement issues, and budgetary limitations. If a manufacturer wanted to determine the average lifetime of its batteries, for instance, the company would definitely not be willing to test every battery produced!

Help students explore situations in which a *sample* would and would not provide good information about a *population*. For example, to estimate the average height of students in their school, students shouldn't use the school's basketball team as a sample. How about the students in a particular English class? Or, if students wanted to find out what percent of students in their school like the food served in the cafeteria, should they send an e-mail to all students in the school and use the first 30 students who respond as the sample?

Discuss the paragraph on "data ethics" with your students.

Are hot dogs unhealthy? What percent of people wear their seat belts when driving? Which works better—a low fat diet or a low carbohydrate diet? Would most teenagers keep an extra $10 they received in incorrect change at a store, or return it? Does listening to music hurt students' concentration and ability to study? How are peoples' heights and foot lengths related? These are just a few examples of the types of questions that statistics can help us answer. Getting clear answers to such questions requires data that have been produced according to a careful plan, as the following example illustrates.

Research question: *Do most people wash their hands after using the bathroom?*

Not according to a December 2005 newspaper article titled "Many Adults Report Not Washing Their Hands When They Should, and More People Claim to Wash Their Hands than Who Actually Do."[1] But before you believe such a headline, you should always ask, "Where did the data come from?"

The article mentioned in the previous paragraph was based on two studies that were done in August 2005. In the first study, 1,013 U.S. adults were asked questions about their hand-washing habits by telephone. This is an example of a **survey**. In the second study, observers watched and recorded the actual hand-washing behaviors of 6,336 adults in public restrooms in four major U.S. cities. This is an example of an **observational study**. Both studies were carried out by Harris Interactive, a company that specializes in these kinds of statistical research.

Now that you know how the data were produced, you might be interested in some results from the two studies.

> While 91% of surveyed adults *claimed* to always wash their hands after using the bathroom, only 83% of the adults in the observational study did so.
>
> In the survey, 94% of women claimed to always wash their hands after using the bathroom, compared with 88% of men. In the observational study, 90% of the women actually washed their hands, compared with 75% of men.
>
> A similar observational study done in 2003 revealed that 78% of the adults observed actually washed their hands after using the bathroom. In that study, 83% of the women and 74% of the men were observed washing their hands.

Based on these studies, what can we conclude? Can we conclude that 83% of *all* U.S. adults always wash their hands after using the bathroom? No, because researchers only observed a **sample** of 6,336 adults, not the entire **population** of U.S. adults. If another group of 6,336 adults was observed on a different day, the percent who washed would probably not be exactly 83%. Can we at least say that the actual percent of all U.S. adults who always wash their hands after using the bathroom is "close" to 83%? That depends on what you mean by "close."

---

1   Harris Interactive, December 14, 2005.

The process of carrying out a statistical study—like the survey or observational study in the previous example—begins with the clear statement of a **research question**. Basically, the research question describes what you want to know in simple terms. Most research questions relate to some population of interest—a group of people, animals, or things. Once a research question has been established, you need to collect some useful data. It's usually not practical to get data from every individual in the population (a **census**). Instead, we usually try to obtain data from a representative sample of individuals chosen from the population. So how do we get the data?

There are three preferred methods for producing data—**observational studies**, **surveys**, and **experiments**. In an experiment, we deliberately do something to one or more groups of individuals—such as giving a drug to people who are sick—and then measure their responses. Observational studies and surveys, on the other hand, attempt to gather data on individuals as they are. In an observational study, we record values of one or more variables—like gender or height—for a sample of individuals. We can obtain these values from direct observation, measurement, or existing data. In a survey, we select a sample of people and have them answer one or more questions. You have already seen examples of a survey and an observational study about people's hand washing habits. How might an experiment shed more light on this subject?

Some people might argue that having an observer present in the restroom—even if the observer isn't washing his or her hands—could influence an individual's hand-washing behaviors. To test this idea, we could design an experiment. Half of the time, we would station an observer at one of the sinks. The other half of the time, we would "hide" the observer in one of the bathroom stalls with a clear view of the sink area. Then, we could compare the percent of people who washed their hands under each of these experimental conditions, called **treatments**.

Each data production method comes with advantages and limitations that you need to understand before you can plan a study. The method used to produce the data also determines the kinds of conclusions that can be drawn. Choosing the best method for a given research question requires careful thought and a lot of practice.

Once we have our data in hand, we must try to figure out what they're telling us. We begin by making graphs and calculating numerical summaries. Then, we interpret the results of our analysis. Of course, our goal is to answer the original research question. Finally, we can communicate our findings to others who might be interested.

Here is a brief outline that summarizes the entire process.

Carrying Out a Statistical Study[2]

    I. Formulate the research question

2   *Guidelines for Assessment and Instruction in Statistics Education (GAISE) Report*, The American Statistical Association, January 2007. *www.amstat.org/education/gaise*

Do some background research to understand the nature of the problem.

Think carefully about what you expect to find and why.

II. Collect data

Decide what to measure and how to obtain the measurements. Which method—survey, observational study, or experiment—would be best?

Think about how you will analyze the data.

Be sure to consider ethical issues.

Produce data according to your stated plan.

III. Analyze the data

Use graphical and numerical summaries to describe the data.

If appropriate, use inference methods to estimate population values or test claims about characteristics of the population.

IV. Interpret your results

Draw conclusions from your data analysis. Remember to answer the research question!

Address any limitations in your conclusions that result from the process of data collection and data analysis.

Communicate your findings.

In this module, you will learn how to analyze surveys, observational studies, and experiments that have been planned by others. Then, you'll get to design and carry out your own studies. As you go, keep this in mind: You can't draw sound conclusions from badly produced data.

Here's another important principle to remember: Statistical studies should be conducted in an ethical manner. Avoid the use of deception whenever possible and ensure that survey participants and experimental subjects are informed about the purpose of the study and any potential risks associated with their participation. If study subjects are people, they must provide their informed consent to participate after being made aware of any potential risks that may result from taking part in the study. For studies involving minors, parent/guardian permission is required. If the study uses animals as subjects, researchers should follow published guidelines for humane treatment of animals, such as those published by the American Psychological Association (see *www.apa.org/science/anguide.html*). Researchers should also ensure the anonymity and confidentiality of peoples' responses and behaviors unless participants have been informed that responses will not be confidential. For experiments, it is common to have a review board approve the experimental design in advance and monitor the results of the experiment as data are collected.

Teacher Notes for Investigation #1: Did You Wash Your Hands?

This investigation builds on the two hand-washing studies that were discussed in the Introduction. The questions posed here are designed to get students thinking about statistics in practice and to provoke discussion in the classroom. We try to alert students to several important issues, such as:

> The way in which data are produced affects the kinds of conclusions that can be drawn. Only well-designed experiments can be used to make cause-and-effect conclusions.

> An estimate we obtain from a sample could differ greatly from the truth about the population if:

>> Our sample doesn't represent the population well

>> The question we asked is unclear or misleading

>> People don't respond accurately or honestly

>> Some people refuse to respond

>> The observer influences the observed

Prerequisites

Students should be able to determine from a narrative description whether data were produced with a survey, an experiment, or an observational study.

Learning Objectives

As a result of completing this investigation, students should be able to:

> Decide which method of producing data—a survey, experiment, or observational study—is most appropriate for answering a given research question

> Describe how certain practical difficulties may affect the results of surveys, experiments, and observational studies

> Define a research question

Teaching Tips

Consider whether you want students to answer the questions in this first investigation individually or with a partner. Either way, be sure to allow time for class discussion of the questions.

Suggested Answers to Questions

Many of the questions do not have "right" or "wrong" answers. Students should be encouraged to defend their answers with specific evidence, much as an attorney would in a legal case.

1. Answers will vary. Students might focus on the unsanitary or disgusting nature of not washing hands after using the bathroom. There are also health-related implications. According to the Harris Interactive news release from the hand-washing study, "Infectious diseases, many caused by unclean hands, are the leading causes of death and disease worldwide and the third leading cause of death in the United States. The Centers for Disease Control and Prevention (CDC) says that hand washing is the single most important means of preventing the spread of infection."

2. (a) Those who do not have telephones. Also, if calls were placed only to landlines, then those who only have cellphones would be left out.

(b) Answers will vary. Here's one possible answer. As adults who do not have telephones would tend to be poor, perhaps even homeless, they might not have access to proper facilities for washing their hands after using the bathroom. Without these people's behaviors represented in the survey, the 91% estimate would be too high.

(c) Answers will vary. Here's one possible answer. Some people would be embarrassed to admit that they don't usually wash their hands after using the bathroom, and so might refuse to answer. Other individuals might feel that they are too busy to participate in a survey.

(d) Since this survey asks about a potentially embarrassing issue, it seems likely that some people will give a socially acceptable "always" answer, even if this is not the truth. It is also possible that some individuals will have inaccurate recollections of their hand-washing habits.

3. (a) Answers will vary. Here's one possible answer. When others are present in the bathroom, some people might be more likely to wash their hands due to implicit "peer pressure." If that's the case, then a hidden camera would have revealed fewer than 83% who washed their hands.

(b) Answers will vary. Here's one possible answer. People may be generally more likely to follow societal expectations to wash their hands when they are in public than when they are at home. If so, then a hidden camera study would reveal less than 83% who washed their hands at home.

4. (a) Answers will vary. Here's one possible answer. From these two studies, it certainly appears that a higher percent of adults claim to wash their hands after using the bathroom than the percent who actually do so when observed. However, the survey and the observational study involved different groups of people. It is possible that the difference in the results of the two studies (91% who claimed they washed their hands versus 83% who actually did) is due to differences in hygiene habits between the people in these two groups, and not from people's tendency to overestimate their hand-washing tendencies. After all, if researchers had observed a different group of 6,336 adults in the same public restrooms on a different day, the percent who were seen washing their hands would probably not have been exactly 83%. Likewise, if researchers had surveyed another sample of 1,013 adults by telephone, it is unlikely that exactly 91% would say they always washed their hands after using the restroom. The difference in the results of these two studies may just be due to the natural variability that occurs from one sample to another.

(b) Answers will vary. Here's one possible answer. Conduct a hidden camera observational study of a sample of individuals in public restrooms. Then survey those same individuals about their hand-washing habits after the bathroom visit.

5. It's always important to know who sponsored (paid for) a statistical study. In this case, the sponsoring agency would likely be hoping to encourage people to do a better job of washing their hands after using the bathroom, thereby using more soap. A study that reveals a lower percent of people washing their hands would suit this agency's purpose.

6. (a) For example, "What percent of teenagers always wash their hands after using the bathroom?"

(b) A survey or an observational study would both be reasonable methods for producing data. An experiment wouldn't be appropriate, as the goal of the study is to record teenagers' normal hand-washing tendencies, not to try to do something to influence those tendencies.

(c) Answers will vary. Here's one possible answer. Teenagers tend to pay less attention to personal hygiene and health than adults do, so the percent of teenagers who always wash their hands after using the bathroom would probably be lower.

7. (a) A survey. Since it would be difficult to actually observe teenagers brushing their teeth, an observational study wouldn't be practical. Because the goal of the study is to record teenagers' normal tooth-brushing habits, not to try to do something to influence those habits, an experiment would not be appropriate. If possible, the teens who participate in the survey should be selected at random from the population of teens in the area. Note that it may be necessary to obtain parental permission before surveying teenagers.

(b) An experiment. If researchers deliberately give drug A to one group of individuals and drug B to another, then they can compare the differences in the percent in each group who experience nausea following their migraine headaches. If the decisions about which individuals get drug A and which drug B are made at random, then researchers can determine whether the difference in the percents who experience nausea is sizable enough to suggest a difference in the effectiveness of drug A and drug B. (By assigning the drugs at random to the migraine sufferers, researchers help ensure that the group of individuals taking drug A and the group taking drug B are fairly balanced in all ways that might affect their response to the drug treatments. If the two groups are similar to begin with, then any sizable differences that emerge between the two groups after the drugs are administered can be attributed to the effects of the drugs themselves.)

Note that a comparative observational study using two groups of people—one group who have used drug A and the other group who have used drug B—would not allow researchers to establish any kind of cause-and-effect relationship between the drug administered and people's tendency to have nausea later. Since people have chosen whether to use drug A or drug B, it is possible that the two groups of individuals differ systematically in other ways that might affect their likelihood of becoming nauseated after having a migraine.

(c) Either an observational study or a survey. Asking a (random) sample of males and a (random) sample of females to report how many numbers are stored in their phones will allow for direct comparison of the average number of contacts for the two genders. However, some individuals may report inaccurate values. Actually observing the phone lists of the randomly selected people might result in more reliable data. An experiment would not be appropriate since we only want to observe what is true, not try to influence the state of affairs.

(d) Observational study. Watching the actual behavior of drivers at the stop sign in question would be more effective than asking them whether they stop. It would be best if observers could watch without being noticed by the drivers, since the presence of an observer may influence the behavior of the driver. An experiment would not be appropriate because we are simply trying to observe and record whether drivers stop completely, not to influence whether they stop.

(e) An experiment, with half of the customers receiving a bill with suggested tip amounts at the bottom and half of the customers receiving no suggested tip amounts. Ideally, the determination of which customers get the bills with suggested tip amounts should be made at random so that the two groups of customers will be as similar as possible in every respect that might influence the amount they decide to tip other than the intended "treatment"—suggested tip amounts on the bill versus no suggested tip amounts on the bill. Then, any substantial difference that emerges between the average tip amounts in the two groups could be attributed to whether suggested tip amounts were printed on the restaurant bill.

8. Answers will vary. The survey results from the two years were very similar—91% of respondents said they always washed their hands after using the bathroom in 2005, compared with 92% in 2007. However, the observational study results from the two years were quite different. In 2005, 83% of those observed washed their hands after using the bathroom. In the 2007 observational study, only 77% washed their hands after using the bathroom. One possible explanation for this decrease is the decline in the percent of men who washed their hands—from 75% in 2005 to 66% in 2007.

Possible Extensions

Ask students to develop a research question that would best be answered by (a) an observational study, (b) a survey, (c) an experiment.

Have students locate an article describing the results of a survey in printed or electronic media. Then ask them to identify the research question, the population, the sample, and any concerns they have about the results reported in the article.

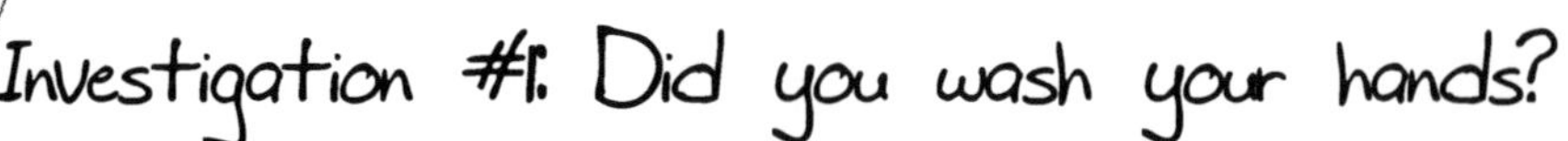 

Investigation #1: Did you wash your hands?

1. Why should we care whether people wash their hands after using the bathroom?

Corresponds to pp. 4-9
in Student Module

2. In the Harris Interactive survey, people were contacted by telephone. One of the questions the interviewers asked was, "How often do you wash your hands after using a public restroom?"

(a) Which U.S. adults were not included in this study?

(b) The survey estimated that 91% of all U.S. adults would claim that they always wash their hands after using the bathroom. Do you think this estimate is too high, too low, or about right given your answer to (a)? Explain.

(c) Several people refused to participate in the survey. Give a reason that this might happen.

(d) In any survey, it is possible that some people will not answer a question accurately or honestly. Thinking about the hand-washing survey, do you think this is likely to happen? Explain your answer.

3. The observational study of hand washing was conducted at a baseball field in Atlanta, a museum and an aquarium in Chicago, a bus and train terminal in New York, and a farmers' market in San Francisco.

(a) Observers in the public bathrooms combed their hair or put on make-up at one of the available sinks while they were watching individuals' hand-washing behaviors. If the observation had been done by hidden camera instead (with no observer present), do you think the percent who washed their hands would have been greater than, less than, or about the same as 83%? Justify your answer.

(b) Suppose the observational study had been conducted using hidden cameras in the homes of the same 6,336 adults. Do you think the percent of these individuals who washed their hands would have been greater than, less than, or about the same as 83%? Justify your answer.

4. (a) Comment on the conclusion reached in the newspaper headline: "More People Claim to Wash Their Hands than Who Actually Do."

(b) Describe a study design involving only one group of people that might help us better evaluate the validity of the quoted claim in part (a).

5. Both studies were paid for by the American Society for Microbiology and the Soap and Detergent Association. Should you take this information into account when interpreting the results of the studies? If so, how?

6. You have been asked to help design a study to investigate how often teenagers wash their hands after using the bathroom.

(a) Define a research question for your study.

(b) Would you recommend using a survey, an observational study, or an experiment to produce the data? Explain.

(c) Do you think the percent of teenagers who always wash their hands after using the bathroom is higher than, lower than, or about the same as the percent of adults who do so? Justify your answer.

7. For each of the following research questions, decide which method of data production—a survey, an experiment, or an observational study—would be most appropriate. Justify your choice of method.

(a) What percent of teenagers leave the water running while they brush their teeth?

(b) Which of two drugs is more effective at preventing nausea following the onset of a migraine headache?

(c) Do male teenagers or female teenagers tend to have more numbers stored in their cell phones?

(d) What percent of drivers come to a complete stop at a stop sign near a local elementary school?

(e) Does printing suggested tip amounts on the bottom of a restaurant bill increase the average amount that customers leave in tips?

8. A follow-up study conducted in 2007 by Harris Interactive revealed that while 92% of adults said that they always washed their hands after using the bathroom, only 77% of the adults observed in public restrooms actually did. According to Harris Interactive's Hand Washing Fact Sheet, "The overall decline in hand washing observations is largely due to males. The percentage of males observed washing their hands fell from 75% in 2005 to 66% in 2007. Overall, the percentage of females observed washing their hands is down slightly from 90% in 2005 to 88% in 2007."

Did people's hand washing habits improve or get worse from 2005 to 2007? Justify your answer with specific evidence from the reports describing the Harris Interactive studies.

Teacher Notes for Section I: Observational Studies

In Section I of the module, students examine observational studies. There are five investigations in this section. The first three involve data that are produced with no random selection. In the last two investigations, the data production process involves random selection. As we discuss in the section overview, random selection helps ensure that a "representative" sample is obtained from some larger population of interest. When random selection is used in an observational study, we can generalize results from the sample to the larger population with confidence. Without random selection, our ability to generalize is limited.

The five investigations in this section are:

Investigation #2: Get Your Hot Dogs Here!

Students analyze nutritional data on different brands and types of hot dogs from an observational study carried out by Consumers Union.

Investigation #3: What's in a Name?

Students examine the popularity of the first names of students in their class.

Investigation #4: If the Shoe Fits …

Students play the role of statistical detective as they try to identify a culprit using data on the foot lengths and heights of a random sample of students at a school.

Investigation #5: Buckle Up

Students explore whether seat belt use among drivers is improving in the states.

Investigation #6: It's Golden (and It's Not Silence)

In this culminating investigation, students design, implement, analyze data from, and draw conclusions from an observational study involving people's preference for the golden ratio.

Prerequisites

Students should be able to:

Distinguish an observational study from a survey or an experiment

Describe the distribution of a categorical variable using counts, percents, and bar graphs

Describe the shape, center, and spread of the distribution of a quantitative variable using a dotplot and numerical summaries (mean, median; range, interquartile range (IQR), standard deviation; five-number summary)

Identify potential outliers

Compare distributions of a quantitative variable using back-to-back stemplots, parallel boxplots, and numerical summaries (mean, median, mode; range, interquartile range (IQR), standard deviation)

Describe the relationship between two quantitative variables using a scatterplot

Examine the effect of adding a categorical variable to a scatterplot on the relationship between two quantitative variables

Use a scatterplot with or without a summary line to make predictions of y from x

Interpret the slope and y-intercept of a summary line in the context of a problem

Learning Objectives

As a result of completing this section, students should be able to:

Explain why random selection in an observational study allows sample results to be generalized to a larger population of interest

Describe the relative standing (percentile or z score) of one value within a distribution

Explain why an observational study might be preferable to a survey in describing individuals' behavior

Construct an appropriate graphical display (dotplot, stemplot, or histogram) of a quantitative variable

Describe how the method of data production affects their ability to draw conclusions from the data

Carry out a complete analysis of an observational study involving one or more categorical variables, using counts, percents, and bar graphs to support their narrative comments

Carry out a complete analysis of an observational study involving one or more quantitative variables, using dotplots, stemplots, boxplots, and numerical measures of center and spread to support their narrative comments

Design a practical sampling plan that incorporates random selection

Conduct an observational study in a way that should produce reliable data

Draw conclusions about a population based on graphical and numerical information from a representative (random) sample

Teaching Tips

Take time to go through each of the examples in the overview for this section. We discuss each one in detail below.

The first example addresses the research question: Who talks more, men or women? Note the individual-to-individual variation that's present in the number of words spoken per day. This interesting (and perhaps surprising) observational study appears to have debunked an earlier claim that women talk about three times as much as men. Be sure to point out one obvious limitation of this study: All the participants were college students from the United States and Mexico. These results may not apply to college students in other countries, or to people of other ages.

In the second example, we consider a quality control setting. This time, the research question is: Do the potato chips produced today meet specifications in terms of their salt content? Of course, we expect some variation in salt content from chip to chip. Based on a representative sample of chips from the day's production, we should be able to draw a conclusion about the salt content in the larger population of chips produced today.

How can we get a "representative" sample of individuals from some population of interest? Using random selection. In an ideal world, we would put a slip of paper representing each individual in the population in a hat, mix the slips thoroughly, and then draw out a sample of the desired size. Random selection entails letting chance decide which individuals from the population of interest end up in a sample.

Using the idealized hat method, every individual in the population has an equal chance of being selected for the sample. In addition, every subgroup of n individuals in the population is equally likely to be chosen as the sample (for any sample size n).

In the third example, a high-school student (Kayla) undertakes a study to answer the research question: What is the average number of contacts stored in seniors' cell phones?

We show the mechanics of using a table of random digits and a random number generator to mimic the hat method for Kayla's study. It is important to stress that before heading for the random digits table or random number generator, students should have already assigned a unique numeric label to each individual in the population.

It is usually not practical to obtain a true random sample. We ask students to think about some reasonable alternatives involving random selection for Kayla's cell phone study and for the potato chip quality control study.

Possible Extension

You might want to introduce students to other methods of sampling that involve random selection, such as stratified sampling, cluster sampling, systematic sampling, and multi-stage sampling. For more information about these sampling methods, consult any AP Statistics textbook.

You can learn a lot just by watching. That's the idea of an observational study. If you want to know how often people wash their hands after using the bathroom, don't ask them! Observe them. As we saw in the Introduction, what people say and what they actually do can be quite different. But be sure to keep in mind the old adage: "The observer influences the observed." Merely having an observer present in the restroom might affect the percent of people who wash their hands.

In her book, *The Female Brain*, Dr. Louann Brizendine claimed that women talk almost three times as much as men. Some researchers at the University of Arizona were skeptical, so they designed an observational study to examine this claim. About 400 male and female college students participated in the study. The students wore specially designed recording equipment that turned on automatically at pre-set intervals over several days without the students' knowledge. Researchers then counted words used by the male and female participants. Their findings? Both males and females tended to speak an average of about 16,000 words per day. Dr. Brizendine later admitted that her claim had little factual basis.

Let's consider one further example from industry. Suppose you are in charge of quality control at a factory that produces potato chips. Imagine a string of thousands of very similar looking chips moving one behind the other down a conveyor belt, hour after hour, day after day. At some point in the process, salt is added to each chip. How can you be sure that the chips your factory is producing today don't contain too much or too little salt? Do you have to measure the salt content of every potato chip made today? Of course not. It isn't practical to observe every chip. Even if it were, you wouldn't choose to do that, because measuring the amount of salt on a chip actually destroys the chip. If you examined the salt content of every chip produced that day, you'd have no potato chips left to sell! What should you do instead? Select a sample of chips from that day's production and measure the salt content of the chips in the sample.

The potato chip example reminds us of an issue that was discussed briefly in the Introduction. If we want to get information about some characteristic of a population, such as the salt content of the potato chips produced today, we often tend to measure that characteristic on a sample of individuals chosen from the population of interest. We'd like to draw conclusions about the population based on results from the sample. To generalize from sample to population in this way, we need to know that the sample is representative of the population as a whole.

Suppose you measured the salt content of the last 100 potato chips produced at the factory today and found that the chips were generally too salty. Should you conclude that the entire batch of chips produced today is too salty? Not necessarily. Something may have happened during the last hour of production that affected the saltiness of the chips made at the end of the day. The last 100 chips produced may not be a representative sample from the population of today's potato chips.

So how do we get a representative sample? If we choose the first 100 potato chips, or the last 100, or even 100 chips "willy-nilly" off the conveyor belt, we may obtain a sample in which the chips tend to be consistently saltier than or less salty than the entire batch of chips produced that day. The best way to avoid this problem is to let chance select the sample. For example, you might choose one time "at random" in each of the 10 hours of production and measure the salt content of the next 10 potato chips that pass a certain point on the conveyor belt at those times. This incorporates **random selection** into the way the sample is chosen.

Random selection involves using some sort of chance process—such as tossing a coin or rolling a die—to determine which individuals in a population are included in a sample. If the individuals are people, one simple method of random selection is to write people's names on identical slips of paper, put the slips of paper in a hat, mix them thoroughly, and then draw out one slip at a time until we have the number of individuals we want for our sample. An alternative would be to give each individual in the population a distinct number and use the "hat method" with this collection of numbers, instead of people's names. Notice that this variation would work just as well if the individuals in the population were animals or things instead of people.

The hat method works fine if the population isn't too large. If there are too many individuals in the population, however, we would need a very big hat and many small slips of paper. In such cases, it would be easier to "pretend" that we're using the hat method, but to choose the numbers in a more efficient (but equivalent) way.

Technology is the answer. Computers and many calculators have the ability to select numbers "at random" within a specified range, just like drawing the numbers out of a hat. These devices can generate many numbers at random in a short period of time.

Many statistics textbooks contain entire pages filled with rows of "random digits"—numbers from 0 to 9 generated at random using technology. Such tables of random digits were especially useful before the invention of graphing calculators. Here are four rows of random digits that might appear in such a table:

```
5 2 7 1 1      3 8 8 8 9      9 3 0 7 4      6 0 2 2 7
4 0 0 1 1      8 5 8 4 8      4 8 7 6 7      5 2 5 7 3
9 5 5 9 2      9 4 0 0 7      6 9 9 7 1      9 1 4 8 1
6 0 7 7 9      5 3 7 9 1      1 7 2 9 7      5 9 3 3 5
```

Now let's consider an example. Kayla wants to conduct an observational study investigating the average number of contacts stored in teenagers' cell phones. She decides to restrict her attention to seniors, most of whom have cell phones. There are 780 seniors in her high school. How might Kayla use random selection to choose a sample of 30 seniors to participate in the cell phone study?

20

It would be tedious to write 780 names on slips of paper, so Kayla decides to pretend that she's using the hat method. After getting an alphabetized list of the school's seniors from the office, Kayla numbers the students from 1 to 780 in alphabetical order. To choose 30 seniors at random, Kayla can then use either a random digits table or a random number generator.

Random digits table: To use a random digits table, Kayla could look at groups of three digits, which could range from 000 to 999. If she lets 001 correspond to student 1 on the list, 002 correspond to student 2, and so forth, then 780 would correspond to student 780, the last senior on the list. Numbers 781, 782, ..., 000 would not correspond to any of the students on the list. By starting at the left-hand side of a row in the table and reading across three digits at a time, Kayla would continue until she had chosen 30 distinct numbers between 001 and 780. The corresponding seniors would be the chosen sample.

Using the lines of random digits on the previous page, for example,

```
5 2 7 1 1       3 8 8 8 9       9 3 0 7 4       6 0 2 2 7
4 0 0 1 1       8 5 8 4 8       4 8 7 6 7       5 2 5 7 3
9 5 5 9 2       9 4 0 0 7       6 9 9 7 1       9 1 4 8 1
6 0 7 7 9       5 3 7 9 1       1 7 2 9 7       5 9 3 3 5
```

the senior numbered 527 would be chosen first, and the senior numbered 113 would be selected second. Kayla would skip the numbers 888 and 993 because they don't correspond to any seniors, and so on. Continuing likewise, the first 10 students in the sample would be the seniors numbered 527, 113, 074, 602, 274, 001, 185, 487, 675, and 257. The eleventh student selected would be the senior numbered 395. Do you see why?

Random number generator: Kayla could also use her calculator or computer to generate a "random integer" from 1 to 780. She would repeat this until she got 30 distinct numbers from 1 to 780. The seniors on the alphabetized list with the corresponding numbers would be the chosen sample.

In this example, Kayla entered the command randInt(1,780) on a TI-84 calculator and pressed ENTER several times to repeat the command. The first ten resulting numbers were 718, 512, 653, 416, 190, 89, 689, 519, 470, and 44. So the seniors with these numbers would be included in her sample.

We used the "random integer generator" at *www.random.org* as an alternative and came up with the numbers here.

If random selection is accomplished by using the hat method or mimicking it with random numbers, the resulting sample is called a **random sample**. To be classified as a random

Random Integer Generator

Here are your random numbers:

741	72	355	297	755
559	398	629	47	310
536	304	752	397	483
388	405	149	634	699
739	152	721	516	640
293	589	714	771	566

sample, the n selected individuals must have been chosen by a method that ensures:

(1) each individual in the population has an equal chance to be included in the sample

(2) each group of n individuals in the population is equally likely to be chosen as the sample

In the cell phone study example, Kayla did obtain a random sample. Once she selected the students for her observational study, it might have been quite difficult for Kayla to locate the 30 seniors who were chosen in a school with so many students, however. For practical reasons, Kayla might have used a method of random selection that didn't result in a truly random sample.

If, for example, the 780 seniors were assigned to 30 homerooms of 26 seniors each based on their last names, Kayla might have decided to select one student at random from each homeroom for her cell phone study. Notice that this alternative method of random selection does give each senior in Kayla's school an equal chance to be included in the sample, but it does not give every group of 30 seniors an equal chance to actually be chosen as the sample. In fact, with this method, the chance of getting a sample with two or more students from the same homeroom is zero!

Think back to the potato chip example for a minute. Can you imagine how difficult it would be to take a random sample from all of the potato chips produced in one day? Just picture someone numbering the individual potato chips for starters! It would be much more feasible to select, say, 10 consecutive potato chips from a particular spot on the conveyor belt by choosing a time at random during each hour of production.

Some observational studies do not use random selection to select the individuals who participate. In the hand-washing study from the Introduction, for example, observers simply watched whoever happened to be in public restrooms at the time. Perhaps the kinds of people who use public restrooms at sporting events, in museums or aquariums, and in train stations have different hand-washing habits than the population of adults as a whole.

The researchers from the University of Arizona used volunteer college students from the United States and Mexico in their observational study of talking patterns by gender. Because of the way in which their sample was chosen, their conclusion about male and female talking tendencies wouldn't necessarily apply to older adults or to college students from other countries. In fact, the results might not even extend to all college students, since some—perhaps those who talk a lot—might have refused to participate in the study. Lack of random selection limits our ability to generalize from the sample to a larger population of interest.

In the investigations that follow, you will learn more about designing and analyzing results from observational studies. You will see firsthand how the presence or absence of random selection affects our ability to generalize.

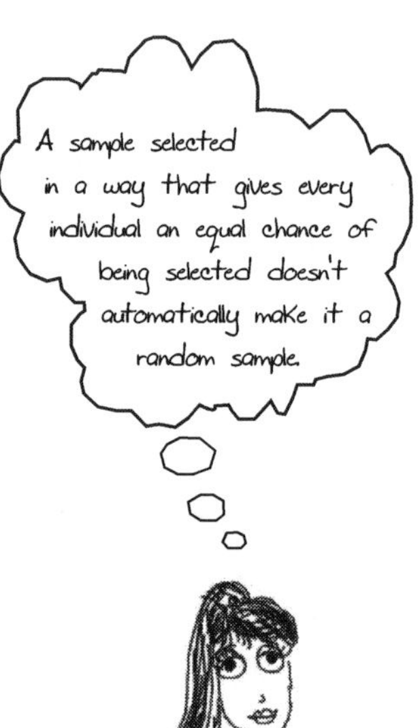

Teacher Notes for Investigation #2: Get Your Hot Dogs Here!

THE FIRST INVESTIGATION IN THE OBSERVATIONAL STUDIES SECTION ASKS STUDENTS TO use graphical and numerical tools of data analysis, combined with a heavy dose of common sense, to examine data from a study that did not involve random selection. In this study, Consumers Union selected a convenience sample consisting of one package each of 54 brands of hot dogs. For each brand, they recorded information on several variables, including calorie content, protein-to-fat rating, sodium content, and an overall sensory rating. In this investigation, students are gradually taken through the process of analyzing the hot dog data collected by Consumers Union. As students answer the questions, they should begin to understand the connection between the way in which the data were produced and the kinds of conclusions that can and cannot be drawn from them.

Prerequisites

Students should be able to:

Distinguish an observational study from a survey or an experiment

Describe the distribution of a categorical variable using counts, percents, and bar graphs

Describe the shape, center, and spread of the distribution of a quantitative variable from a dotplot

Compare distributions of a quantitative variable using back-to-back stemplots, parallel boxplots, and numerical summaries (mean, median, mode; range, interquartile range (IQR), standard deviation)

Describe the relationship between two quantitative variables using a scatterplot

Examine the effect of adding a categorical variable to a scatterplot on the relationship between two quantitative variables

Interpret the slope and y-intercept of a summary line in the context of a problem

Use a scatterplot with a summary line to make predictions of y from x

Explain why lack of random selection limits our ability to generalize sample results to a larger population of interest

Learning Objectives

As a result of completing this investigation, students should be able to:

Describe how the method of data production affects their ability to draw conclusions from the data

Carry out a complete analysis of an observational study involving one or more categorical variables, using counts, percents, and bar graphs to support their narrative comments

Carry out a complete analysis of an observational study involving one or more quantitative variables, using dotplots, stemplots, boxplots, and numerical measures of center and spread to support their narrative comments

Teaching Tips

This investigation is designed to review the essential graphical and numerical tools for describing distributions of categorical and quantitative variables, and for describing relationships between two or more variables. Depending on your students' background with techniques of data analysis, you can choose to spend more time on questions involving methods that are less familiar to them. The chart below summarizes the exploratory data analysis tools that will be required in this investigation.

Setting	Graphs	Numerical Summaries
Categorical variables	Bar graphs, pie charts	Counts, percents, proportions
Quantitative variables	Dotplots, stemplots, histograms, boxplots	Center: Mean, median; Spread: range, standard deviation, interquartile range (IQR)
Relationships between categorical variables	Comparative bar graphs	Two-way tables; comparisons of counts, percents, proportions
Relationships between quantitative variables	Scatterplots	Means and standard deviations; correlation

Here is a question-by-question breakdown of the investigation:

Questions 1 through 6 focus on the data production process.

In questions 7 through 10, students use counts, percents, and different types of bar graphs to analyze the protein-to-fat rating and its relationship to the type of hot dog.

Questions 11 through 14 examine data on calories per frank. Students are asked to use graphical displays—dotplots, stemplots, and boxplots—and numerical summaries to describe calories per frank for the three types of hot dogs.

In questions 15 and 16, students examine the relationship between sodium content and calories per frank using scatterplots, correlation, and summary lines.

Question 17 asks students to acknowledge that lack of random selection limits their ability to generalize conclusions from the sample of hot dogs that was tested by Consumers Union.

The final two questions (18 and 19) allow students to demonstrate their understanding of data analysis techniques for categorical and quantitative variables by looking at sensory rating (includes taste) and sodium content. These two questions could be used for homework or as an alternative assessment.

Suggested Answers to Questions

1. Consumers Union carried out an observational study. They did not ask the hot dogs any questions and they did not deliberately do anything to the hot dogs in this study in order to measure a response.

2. We would expect some variation in the calorie content of individual Oscar Mayer beef frankfurters. Perhaps 148 calories is a typical or average amount of calories in hot dogs of this brand.

3. By following a standard preparation method, Consumers Union attempted to ensure that any differences in sensory qualities identified by the raters were due to the difference in brand of hot dog, and not to any difference in the way the hot dogs were cooked.

4. Because individual hot dogs vary, it would have been better to get an average sensory rating for each brand.

5. Probably not. There may be some reason that these packages were easy to reach in the store—because they were recently delivered, or perhaps because they are nearing their sell-by date. In either case, the conveniently chosen single package of hot dogs of a given brand may not accurately represent the characteristics of all packages of hot dogs of that brand at the store in question, much less in the population of all packages of hot dogs of that brand.

6. Since Consumers Union obtained its random sample from the packages of Armour beef hot dogs that were in this store at the time, they would really only be safe generalizing to the population of packages of Armour beef hot dogs in this store. It may be that all of these packages of hot dogs came on the same truck from the same warehouse or factory and the same batch of production, or they could have come from multiple shipments, factories, or batches.

7. (a) 5/15, or about 33% (b) 5/17, or about 29.4%

8. Most of the protein-to-fat rating values for the beef hot dogs are poor or below average.

9. (a) The graph on the left is a side-by-side bar graph comparing the counts of beef and meat hot dog brands falling into each of the protein-to-fat rating categories. The graph on the right is a segmented bar graph showing the distribution of the percent of each type of hot dog having poor, below average, and average protein-to-fat ratings. Because the number of hot dog brands of each type in the study are not equal (beef = 20; meat = 17), it would probably be better to use the segmented bar graph for comparing the protein-to-fat ratings. To see why, note that the same number of brands of beef and meat hot dogs earned poor protein-to-fat rating scores, but that about 10 percent more meat hot dog brands received poor scores.

(b) If we look only at the percent of brands with poor protein-to-fat ratings, then meat hot dogs are somewhat less healthy. If we compare the combined percents of hot dog brands with poor and below average ratios; however, beef hot dogs are worse by about 10 percent. Neither type of hot dog is very healthy when it comes to protein-to-fat rating.

10. (a) The graph on the left shows the distribution of type of hot dog for each protein-to-fat rating category. For example, we see that of hot dogs with poor protein-to-fat ratings, 50% are beef and 50% are meat. The graph on the right shows the distribution of protein-to-fat rating for each type of hot dog. For example, we see that for beef hot dogs, about 60% of brands have poor protein-to-fat ratings, about 37% have below average protein-to-fat ratings, and about 3% have average protein-to-fat ratings. The graph on the right allows for better comparison of the protein-to-fat ratings among the different types of hot dogs.

(b) Poultry hot dogs are healthiest in terms of protein-to-fat ratings. If we combine the poor and below average ratings, we find 97% of beef hot dog brands, 88% of meat hot dog brands, but only 30% of poultry hot dog brands. In addition, about 35% of the poultry hot dog brands have above average or excellent protein-to-fat ratings.

11. (a) You might have conjectured that the three distinct clusters correspond to the three different types of hot dogs—beef, meat, and poultry. That's really not the case, however. It is certainly true that more of the low-calorie hot dog brands are poultry.

(b) The marked point is for Best's kosher beef lower fat.

12. The dotplot shows two or three distinct clusters of calorie content for the beef hot dogs, with one unusually low value. A "typical" value for a beef hot dog appears to be around 150 calories. There is a lot of variability in the calorie content of beef hot dogs in the study—from about 110 to 190 calories per frank.

13. (a)

```
     Meat            Beef

        7  |10|
           |11|  1
           |12|
 9 8 6 5  |13|  1  2  5  9
   7 6 0  |14|  1  8  9  9        Key:  13|  1 means this brand
       3  |15|  2  3  7  8              advertises 131 calories per frank
           |16|
 9 5 3 2  |17|  5  6
       2  |18|  1  4  6
   5 1 0  |19|  0  0
```

(b) Both distributions appear to have two distinct clusters of calorie content values— one in the 170s–190s and the other in the 130s–150s—and one brand with noticeably lower calorie content. A "typical" value for a meat hot dog is about 153 calories (the median) and a "typical" value for a beef hot dog is about 152.5 calories (the median). The spread of calorie content values is slightly higher for meat hot dogs—the range is 88 for the meat hot dogs and 79 for the beef hot dogs; the interquartile range (IQR) is

42 for the meat hot dogs and 38.5 for the beef hot dogs. Eat Slim Veal brand hot dogs have much lower calorie content (107) than the other brands of meat hot dogs. Best's kosher beef lower fat brand hot dogs have the lowest calorie content (111) among beef hot dogs.

14. (a) The dotplot shows the calorie content for each brand of hot dog, while the boxplot only shows summary values—minimum, first quartile (Q1), median, third quartile (Q3), and maximum.

(b) The boxplot makes it visually easier to compare the center and spread of calorie content for the three types of hot dogs.

(c) From the comparative boxplot, we see that beef and meat hot dogs have similar calorie content, while poultry hot dogs have fewer calories, on average. In fact, 75% of the poultry hot dog brands have fewer calories than the median calorie content (around 153) of beef and meat hot dogs. The spread (variability) in calorie content for each of the three types is fairly similar, as seen by the nearly equal ranges (between 85 and 90) and interquartile ranges (between 40 and 43). From the comparative dotplot, we see that for each type of hot dog, there seem to be two distinct clusters of brands with respect to calorie content, along with one unusual value. For the poultry hot dogs, the unusual value is the maximum (Foster Farms Jumbo Chicken, which has 170 calories). For the beef and meat hot dogs, the unusual value is the minimum.

15. (a) There appears to be a weak, positive relationship between the amount of sodium and the calorie content of hot dogs. That is, hot dogs with more sodium tend to have more calories, and hot dogs with less sodium tend to have fewer calories.

(b) The highlighted point corresponds to a brand of hot dog with a relatively large amount of sodium (more than 500 mg) per frank, but a relatively low amount of calories (around 100) per frank. (This point corresponds to Kroger Turkey brand hot dogs.)

(c) There appears to be a much stronger positive relationship between sodium per frank and calories per frank for meat and beef hot dogs. Most of the points for poultry hot dogs are below the points for beef and meat hot dogs, which reaffirms that poultry hot dogs tend to have fewer calories than the other two types. It also appears that none of the poultry hot dog brands fall among the varieties of hot dog with the lowest amount of sodium.

16. (a) Considering all the brands tested by Consumers Union, the average calorie content per frank is about 147, and the average sodium content per frank is about 425 mg. Overall, the correlation of 0.52 confirms our visual impression of a moderate, positive, somewhat linear relationship between sodium and calories for these brands of hot dogs.

(b) The slope, 0.196, tells us that for each additional mg of sodium a brand of beef hot dog has, our summary line predicts an average increase in calorie content of about 0.2 calories per frank. The y-intercept, 78, means the summary line would predict 78 calories per frank for a brand of beef hot dog with 0 mg of sodium per frank. As all the brands of beef hot dogs tested by Consumers Union had around 300 mg of sodium per frank or more, it is not reasonable to trust such a prediction made with the summary line.

(c) For beef hot dogs: $78 + 0.196(300) = 136.8$ or about 137 calories

For meat hot dogs: $62 + 0.232(300) = 131.6$ or about 132 calories

For poultry hot dogs: $24 + 0.214(300) = 88.2$ or about 88 calories

(d) We would expect our prediction for poultry hot dogs to be least accurate, as those points vary the most around the corresponding summary line. We would expect our prediction for beef hot dogs to be most accurate, as those points vary the least around the corresponding summary line.

17. No. Consumers Union did not use random selection to choose the brands of beef, meat, and poultry hot dogs they tested. As a result, the chosen brands of each type of hot dog may not represent the population of all brands of that type.

18. Because the number of brands tested differs for the three types of hot dogs, it is more appropriate to compare percents or proportions, rather than counts. In the table below, we have converted the original count data to show the percent of brands in each category with above average, average, and below average sensory ratings.

	Sensory Rating		
Type of Hot Dog	Above Avg.	Average	Below Avg.
Beef	15.0%	80.0%	5.0%
Meat	35.3%	47.1%	17.6%
Poultry	5.9%	88.2%	5.9%

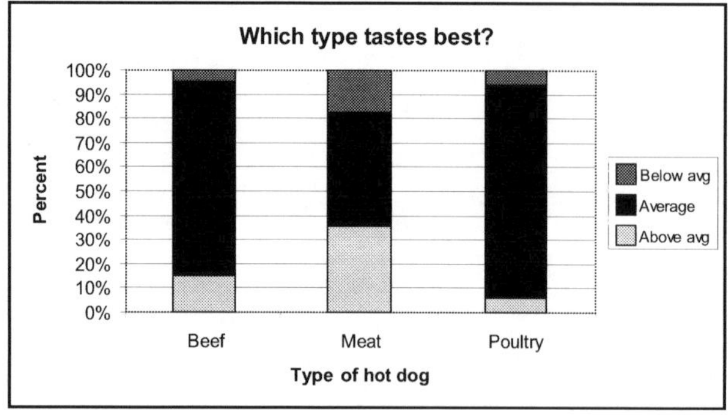

These distributions of sensory ratings are displayed in the bar graph to the left.

A higher percent of meat hot dog brands (35.3%) received above average sensory ratings than for either of the other two types of hot dogs. On the other hand, meat hot dog brands received more below average sensory ratings than did brands of beef and poultry hot dogs. Beef hot dog brands received the smallest percent (5%) of below average ratings, and a respectable percent (15%) of above average ratings. Poultry hot dog brands were generally rated as average in terms of their sensory qualities, including taste.

For the brands tested, if you want to get the best possible sensory experience, and you're willing to take some risk of having an unpleasant sensory experience, then meat brand hot dogs may be the way to go. If you want to have at least an average sensory experience, go with a brand of beef hot dogs. As Consumers Union did not use random selection to choose packages of hot dogs for testing, we should be hesitant to generalize these results to the larger populations of beef, meat, and poultry hot dogs.

19. Comparative graphs and numerical summaries of the sodium content for the three types of hot dogs are shown to the right and below.

From the comparative dotplot, we see that the distribution of sodium content for the tested brands of beef hot dogs has two distinct peaks—one at around 300 mg of sodium per frank and the other at around 475 mg of sodium per frank. For the tested brands of meat hot dogs, the dotplot also shows two peaks—one at around 400 mg of sodium per frank and one at around 500 mg per frank. One brand of meat hot dog, Eat Slim Veal, had unusually low sodium content (144 mg per frank). We can see this potential outlier clearly on the comparative boxplots. There appear to be two distinct clusters of poultry hot dog brands—those with sodium content between 350 and 450 mg per frank and those with sodium content of between 500 and 600 mg per frank.

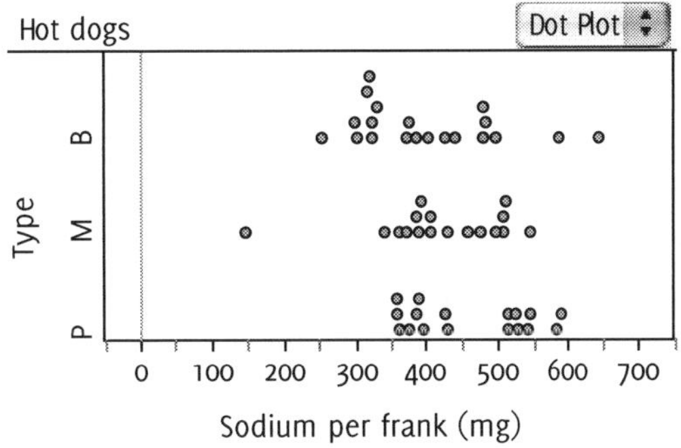

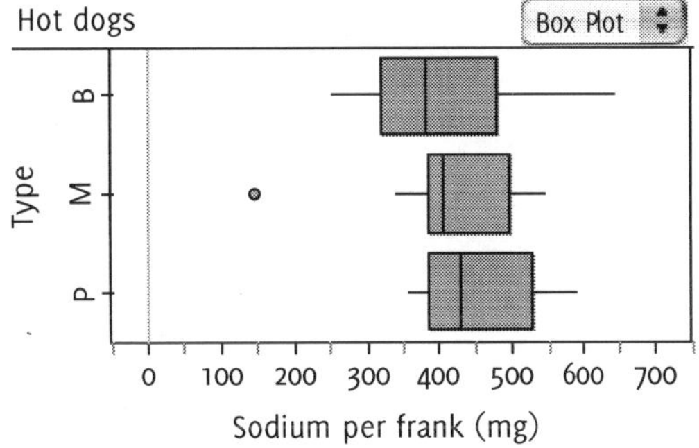

Descriptive Statistics: Sodium per Frank (mg)

Variable	Type	N	Mean	SE Mean	StDev	Minimum	Q1
Sodium	B	20	401.2	22.9	102.4	253.0	319.8
	M	17	418.5	22.8	93.9	144.0	379.0
	P	17	459.0	20.6	84.7	357.0	379.0

Variable	Type	Median	Q3	Maximum
Sodium	B	380.5	478.5	645.0
	M	405.0	501.0	545.0
	P	430.0	535.0	588.0

We see from the boxplots and the numerical summaries (previous page) that poultry hot dogs have the highest median and mean sodium content (430 mg and 459 mg, respectively). Since the first quartile of the poultry and meat brand boxplots is at about the median for the beef brand boxplot, we see that about 75% of the poultry and meat brand hot dogs have higher sodium content than the median (380.5 mg) for beef hot dog brands. Notice that the potential outlier pulls the mean calorie content of the meat hot dog brands well below the mean for the poultry brand hot dogs, even though their two medians are much closer. Beef hot dog brands showed the most variability in sodium content, as we can see from the width of the boxes themselves (interquartile range), and the larger range (around 400 mg). Poultry brand hot dogs have more variability in the middle 50% of the distribution than meat brand hot dogs. The standard deviation of calorie content is higher for meat brand hot dogs due to the large distance of the unusually low value from the mean.

Because Consumers Union did not use random selection to choose packages of hot dogs for testing, we should be hesitant to generalize these results to the larger populations of beef, meat, and poultry hot dogs.

Possible Extensions

How healthy is fast food? Fast food companies often make nutritional data on the products they serve available online or in print. Students could use this available data to compare calories, fat, sodium, and other variables across different companies or different food categories (burgers, chicken sandwiches, etc.). As a starting point, we were able to access McDonald's nutrition facts online at *www.mcdonalds.com/usa/eat/nutrition_info.html*. For Burger King, start at *www.bk.com*. For Wendy's, try *www.wendys.com*.

 # Investigation #2: Get Your Hot Dogs Here!

If baseball is America's game, then hot dogs are America's food. Whether you are at a sporting event, a backyard barbecue, or even a local convenience store, you are bound to see folks wolfing down frankfurters. Why do so many people like to eat hot dogs? For the yummy taste, of course! But what makes hot dogs taste so good? Unfortunately for health-conscious eaters, it's probably the fat and sodium they contain. Not all hot dogs are created equal, however. Some are made from beef, others from poultry, and still others from a combination of meats. With so many varieties available, can hot dog lovers find a healthy option that still tastes great?

Corresponds to pp. 14-28 in Student Module

Several years ago, Consumers Union, an independent nonprofit organization, tested 54 brands of beef, meat, and poultry hot dogs. For each brand tested, they recorded calories, sodium, cost per ounce, a protein-to-fat rating, and an overall sensory rating that included taste, texture, and appearance. The table below and those on the following two pages summarize some of their findings, which were published in *Consumer Reports*.[1] Note that the hot dogs are categorized by type—meat, beef, and poultry.

Meat Hot Dogs				
Brand	**Protein-to-Fat**	**Calories per Frank**	**Sodium per Frank (mg)**	**Overall Sensory Rating**
Armour Hot Dogs	Poor	146	387	Average
Ball Park	Poor	182	473	Above Avg.
Bryan Juicy Jumbos	Poor	175	507	Average
Eat Slim Veal	Average	107	144	Average
Eckrich Jumbo	Poor	179	405	Average
Eckrich Lean Supreme Jumbo	Average	136	393	Average
Farmer John Wieners	Below Avg.	139	386	Average
Hormel 8 Big	Below Avg.	173	458	Above Avg.
Hygrade's Hot Dogs	Poor	195	511	Average
John Morrell	Poor	153	372	Average
Kahn's Jumbo	Poor	191	506	Above Avg.
Kroger Jumbo Dinner	Poor	190	545	Above Avg.
Oscar Mayer Wieners	Poor	147	360	Above Avg.
Safeway Our Premium	Below Avg.	172	496	Above Avg.
Scotch Buy with Chicken & Beef	Poor	135	405	Below Avg.
Smok-A-Roma Natural Smoke	Poor	138	339	Below Avg.
Wilson	Poor	140	428	Below Avg.

---

1   "Hot dogs: There's not much good about them except the way they taste," *Consumer Reports*, June 1986.

Beef Hot Dogs

Brand	Protein-to-Fat	Calories per Frank	Sodium per Frank (mg)	Overall Sensory Rating
A & P Skinless Beef	Poor	157	440	Average
Armour Beef Hot Dogs	Poor	149	319	Average
Best's Kosher Beef	Below Avg.	131	317	Average
Best's Kosher Beef Lower Fat	Average	111	300	Average
Eckrich Beef	Poor	149	322	Average
Hebrew National Kosher Beef	Poor	152	330	Average
Hygrade's Beef	Poor	190	645	Average
John Morrell Jumbo Beef	Poor	184	482	Average
Kahn's Jumbo Beef	Poor	175	479	Average
Kroger Jumbo Dinner Beef	Poor	190	587	Average
Mogen David Kosher Skinless Beef	Below Avg.	139	322	Average
Nathan's Famous Skinless Beef	Below Avg.	181	477	Above Avg.
Oscar Mayer Beef	Poor	148	375	Average
Safeway Our Premium Beef	Poor	176	425	Above Avg.
Shofar Kosher Beef	Below Avg.	158	370	Average
Sinai 48 Kosher Beef	Below Avg.	132	253	Below Avg.
Smok-A-Roma Natural Smoke	Below Avg.	141	386	Average
Thorn Apple Valley Brand	Poor	186	495	Above Avg.
Vienna Beef	Below Avg.	135	298	Average
Wilson Beef	Poor	153	401	Average

Poultry Hot Dogs

Brand	Protein-to-Fat	Calories per Frank	Sodium per Frank (mg)	Overall Sensory Rating
Foster Farms Jumbo Chicken	Below Avg.	170	528	Average
Gwaltney's Great Dogs Chicken	Below Avg.	152	588	Average
Holly Farms 8 Chicken	Below Avg.	146	522	Average
Hygrade's Grillmaster Chicken	Average	142	513	Average
Kroger Turkey	Excellent	102	542	Average
Longacre Family Chicken	Above Avg.	135	426	Average
Longacre Family Turkey	Above Avg.	94	387	Average
Louis Rich Turkey	Average	106	383	Average
Manor House Chicken (Safeway)	Average	86	358	Average
Manor House Turkey (Safeway)	Excellent	113	513	Average
Mr. Turkey	Average	102	396	Average
Perdue Chicken	Average	143	581	Average
Shenandoah Turkey Lower Fat	Above Avg.	99	357	Average
Shorgood Chicken	Below Avg.	132	375	Average
Tyson Butcher's Best Chicken	Average	144	545	Below Avg.
Weaver Chicken	Below Avg.	129	430	Above Avg.
Weight Watchers Turkey	Excellent	87	359	Average

The *Consumer Reports* article did not provide many details about how the hot dog data were produced. Our best guess is that Consumers Union first obtained one package of each of the 54 brands of hot dogs they intended to test. For each brand, they could then determine the protein-to-fat rating and the calories and sodium per frank from information provided on the package. To prepare the hot dogs for taste testing, Consumers Union cooked each frankfurter in boiling water.

1. Did Consumers Union produce these data using a survey, an experiment, or an observational study? Justify your answer.

2. According to the data table, Oscar Mayer beef hot dogs have 148 calories per frank. Does this mean that *every* Oscar Mayer beef hot dog has exactly 148 calories, or is there some variability in calorie count from frank to frank? Explain.

3. Why didn't Consumers Union cook some hot dogs in the microwave, others on a grill, and the rest in boiling water?

4. For the taste testing, would it have been better to rate one hot dog of each brand, or to get an average sensory rating for several hot dogs of each brand? Why?

5. It is possible that someone from Consumers Union went to one grocery store in a particular city and picked up one easy-to-reach packet of each brand of hot dogs. Would this **convenience sampling** method result in a representative sample of each brand of hot dogs? Why or why not?

6. Suppose Consumers Union had used random selection to choose a package of Armour beef hot dogs from a single grocery store for testing. If they obtained an average sensory rating for all the hot dogs in the selected package, to what population could they generalize their results—all Armour beef hot dogs ever produced, all Armour beef hot dogs that have ever been sent to this store, or all Armour beef hot dogs in this store at the time the sample was chosen? Justify your answer.

In this study, Consumers Union recorded several variables for each brand of hot dog, including type of hot dog, protein-to-fat rating, calories, sodium, and sensory rating. Two of these are **quantitative variables**—calories and sodium. Type of hot dog, protein-to-fat rating, and sensory rating are **categorical variables**. When we analyze data, the types of graphs and numerical summaries we should use are determined by the type of data we are analyzing. We begin by examining two of the categorical variables: type of hot dog and protein-to-fat rating.

7. Here is a **two-way table** that summarizes the protein-to-fat ratings by type of hot dog.

Protein-to-Fat Rating	Type of Hot Dog		
	Beef	Meat	Poultry
Poor	12	12	0
Below Avg.	7	3	5
Average	1	2	6
Above Avg.	0	0	3
Excellent	0	0	3

(a) What percent of hot dogs with a below average protein-to-fat rating were made from poultry?

(b) What percent of poultry hot dogs had below average protein-to-fat ratings?

8. Here is an Excel bar graph of the protein-to-fat rating data for the beef hot dogs.

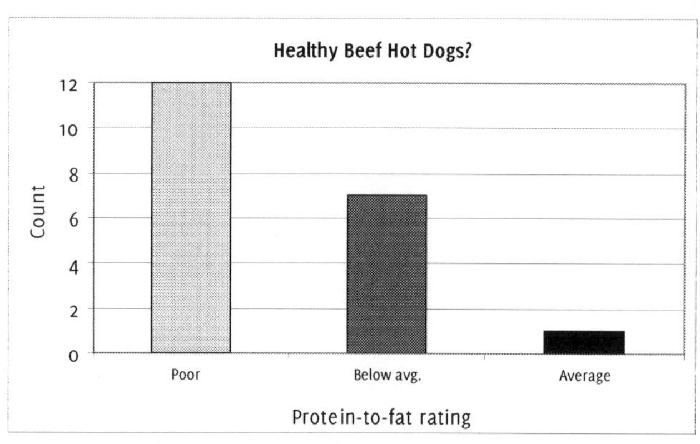

Describe what the graph tells you about protein-to-fat ratings in beef hot dogs.

9. Two Excel bar graphs that could be used for comparing the protein-to-fat ratings for beef and meat hot dogs are displayed below.

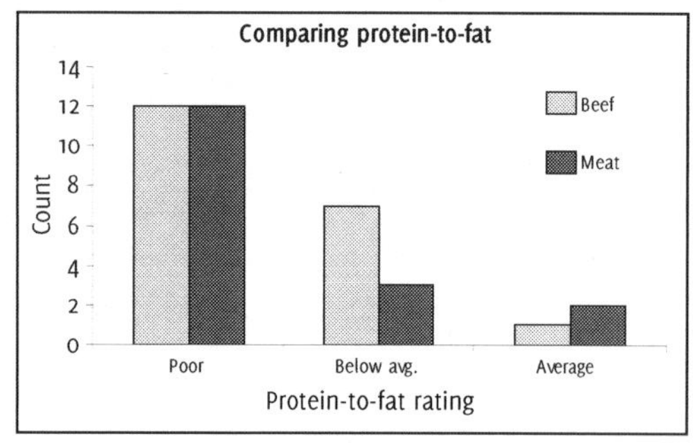

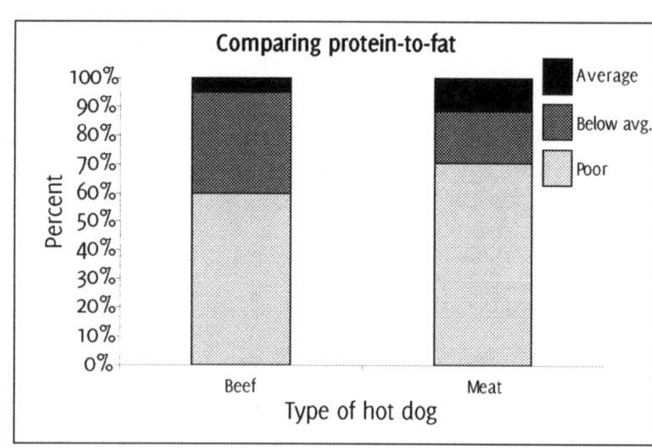

(a) Which graph is more appropriate for making this comparison? Explain.

(b) Write a few sentences comparing protein-to-fat ratings for beef and meat hot dogs.

10. Two different bar graphs that could be used for comparing the protein-to-fat ratings for all three types of hot dogs are displayed below.

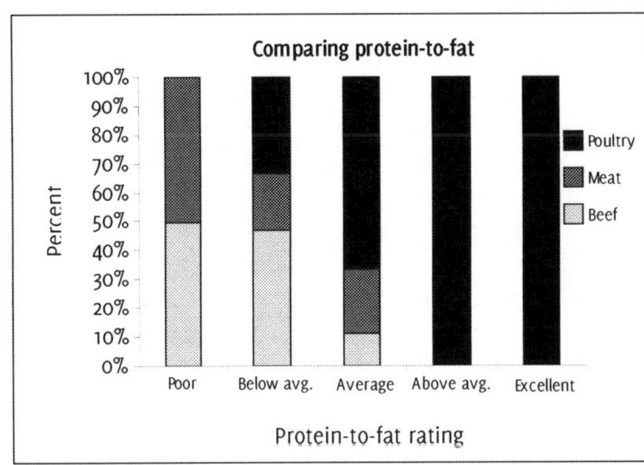

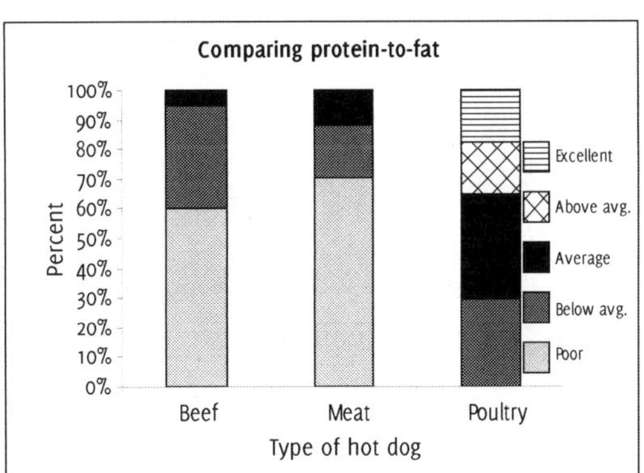

(a) Which graph is more appropriate for making this comparison? Explain.

(b) In terms of protein-to-fat ratings, which type of hot dogs is healthiest? Justify your answer with appropriate graphical and numerical evidence.

Now let's look at the calorie content for different brands of hot dogs.

11. A dotplot of the calorie data for all 54 brands of hot dogs is shown below.

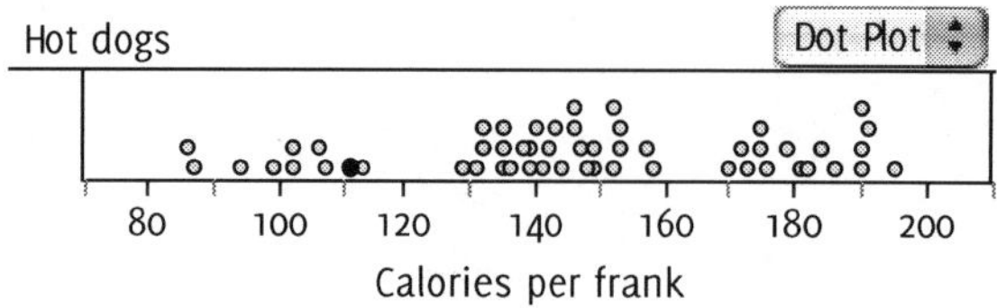

(a) Why do you think this distribution has three distinct clusters? Check whether your hunch is accurate.

(b) Identify the brand and type of hot dog for the highlighted point.

12. A dotplot of the calorie content for the 20 brands of beef hot dogs is shown below. Describe the interesting features of this distribution.

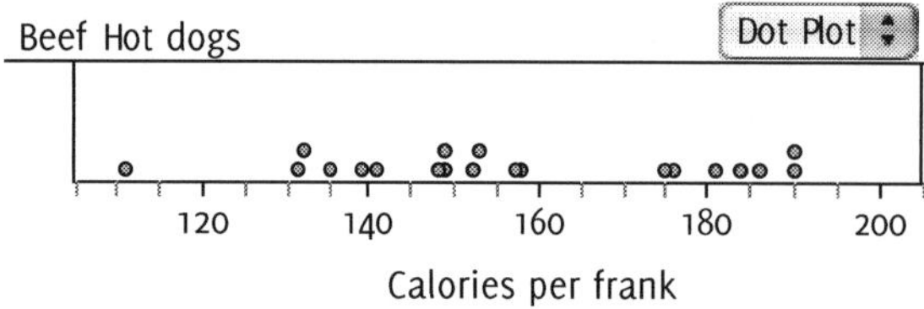

13. How does the calorie content of beef and meat hot dogs compare? A partially completed back-to-back stemplot of the calorie data for these two types of hot dogs is shown below.

```
Meat              Beef

          |10|
          |11|  1
          |12|
          |13|  1  2  5  9        Key: 13|  1 means this brand
          |14|  1  8  9  9        advertises 131 calories per frank
          |15|  2  3  7  8
          |16|
          |17|  5  6
          |18|  1  4  6
          |19|  0  0
```

(a) Add the calorie data for the meat hot dogs to the stemplot. Note that in a back-to-back stemplot, the "leaves" increase in value as you move away from the "stem" in the center of the graph.

(b) Comment on any similarities and differences in the distributions of calories per frank for these two types of hot dogs. Be sure to address center, shape, and spread, as well as any unusual values.

14. To compare calories per frank for all three types of hot dogs, we used computer software to construct graphs and numerical summaries.

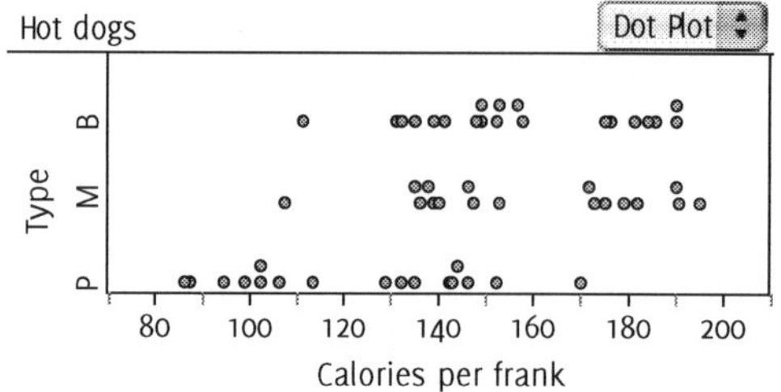 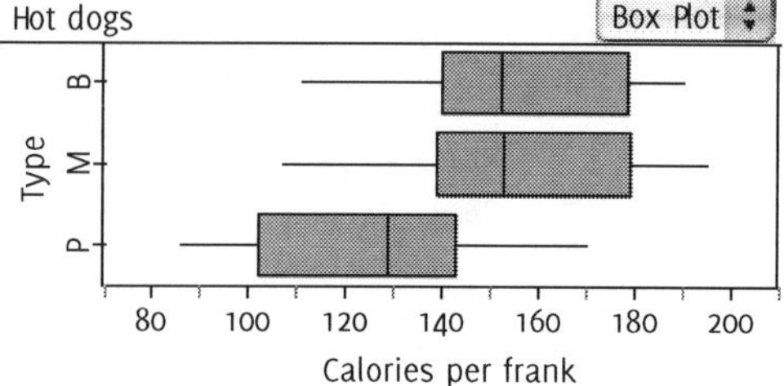

Descriptive Statistics: Calories per Frank by Type

Variable	Type	N	Mean	Median	TrMean	StDev
Calories	B	20	156.85	152.50	157.56	22.64
	M	17	158.71	153.00	159.73	25.24
	P	17	122.47	129.00	121.73	25.48
Variable	Type	SE Mean	Minimum	Maximum	Q1	Q3
Calories	B	5.06	111.00	190.00	139.50	179.75
	M	6.12	107.00	195.00	138.50	180.50
	P	6.18	86.00	170.00	100.50	143.50

(a) Describe one advantage of using the dotplot instead of the boxplot to display these data.

(b) Describe one advantage of using the boxplot instead of the dotplot to display these data.

(c) How do beef, meat, and poultry hot dogs compare in terms of calorie content? Justify your answer using appropriate graphical and numerical information.

Research Question: Is there a relationship between the calorie content and the amount of sodium per frank in these brands of hot dogs?

15. The scatterplot below summarizes the sodium and calorie data for the 54 brands of hot dogs in the Consumers Union study.

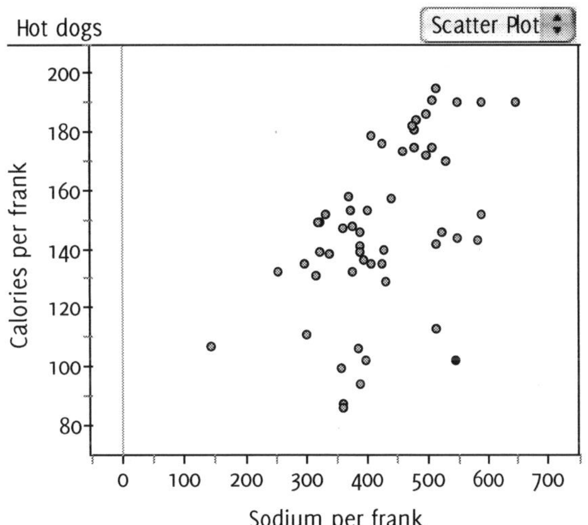

(a) Describe any interesting features of the scatterplot in the context of this study.

(b) What is unusual about the highlighted point in the scatterplot on the previous page?

Here is another scatterplot of the sodium and calorie data with the type of hot dog identified.

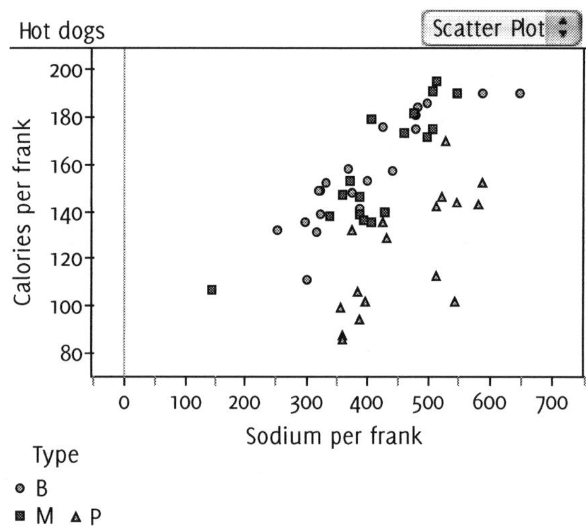

(c) What more can you say about the relationship between sodium and calories per frank when type of hot dog is considered?

16. The next two displays show some numerical summaries of the calorie and sodium data.

Hot dogs

	Calories per frank	Sodium per frank (mg)
	146.611	424.833
	29.0773	95.8564

S₁ = mean ( )
S₂ = stdDev ( )

Hot dogs

	Calories per frank
Sodium per frank (mg)	0.516054

S₁ = correlation ( )

(a) What additional information about the relationship between sodium and calorie content of hot dogs do these numerical summaries provide?

The graph below includes three summary lines—one describing the relationship for each type of hot dog.

(b) Interpret the slope and the y-intercept of the summary line for beef hot dogs.

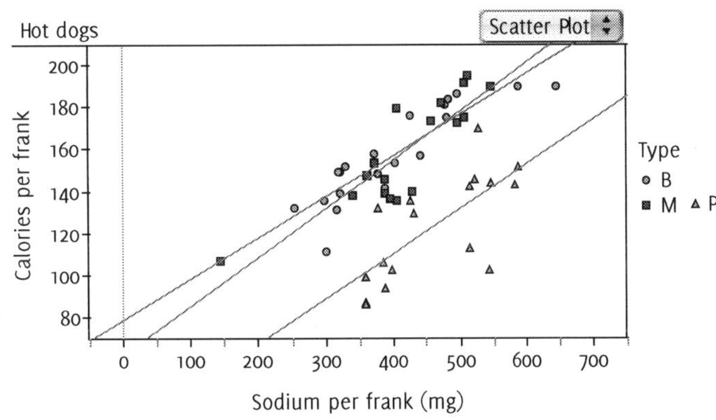

— ○ Calories per frank = 78 + 0.196Sodium per frank (mg); r² = 0.79
— ■ Calories per frank = 62 + 0.232Sodium per frank (mg); r² = 0.75
— ▲ Calories per frank = 24 + 0.214Sodium per frank (mg); r² = 0.51

(c) Suppose Consumers Union had chosen another brand of meat hot dog, beef hot dog, and poultry hot dog, each having 300 milligrams of sodium per frank. What would you predict for the calories per frank in each case? Explain how you made your prediction.

(d) Based on the graph on the previous page, which of the predictions in the previous question do you think would be most accurate? Explain.

17. In the Consumers Union study, beef hot dogs had a mean calorie content of 156.85 calories per frank, compared to 158.71 calories per frank for meat hot dogs and 122.47 calories per frank for poultry hot dogs. Would you feel comfortable generalizing this result about calorie content to the *population* of all brands of beef, meat, and poultry hot dogs? Why or why not?

18. What about the taste? Consumers Union gave an overall sensory rating, which included texture, taste, and appearance. The following table summarizes the ratings by type of hot dog.

Type of Hot Dog	Sensory Rating		
		Above Avg.	Average	Below Avg.
	Beef	3	16	1
	Meat	6	8	3
	Poultry	1	15	1

Which type of hot dog had the best overall sensory ratings? Prepare a brief report that includes graphical and numerical evidence to support your answer.

19. How salty are they? Which have more sodium per frank—beef, meat, or poultry hot dogs? Carry out an analysis that includes graphs and numerical summaries to help answer this question. Write a brief report that summarizes your analysis on a separate piece of paper.

Teacher Notes for Investigation #3: What's in a Name?

This second investigation in the Observational Studies section presents students with another example of a study in which no random selection is used. In the previous investigation, students used data that had been produced by someone else. This time, they will collect their own data on the popularity of the first names of students in their class. Then, students will analyze the data with appropriate graphs and numerical summaries. Finally, students are asked to reflect on how the method of data production affects their ability to generalize results.

Prerequisites

Students should be able to:

> Use proportions to answer questions involving categorical variables

> Describe the shape, center, and spread of the distribution of a quantitative variable

> Distinguish an observational study from an experiment

> Explain why lack of random selection limits our ability to generalize sample results to a larger population of interest

Learning Objectives

As a result of completing this investigation, students should be able to:

> Construct an appropriate graphical display (dotplot, stemplot, or histogram) of a quantitative variable

> Describe how the method of data production affects their ability to draw conclusions from the data

Teaching Tips

This is a fairly straightforward observational study based on research showing that people's first names can affect their lives in unexpected ways. For example, two economics professors conducted a study titled "Are Emily and Brendan More Employable Than Lakisha and Jamal? A Field Experiment on Labor Market Discrimination."[1] These researchers constructed sets of resumes for fictitious individuals, some of whom were highly qualified and others of whom were less qualified. For each resume, they gave the fictitious applicant either a "white sounding" or "black sounding" name. Then, they sent about 5,000 resumes out in response to more than 1,000 jobs that were advertised in Chicago and Boston. What did they find? Candidates with "white sounding" names were 50% more likely to be called back for an interview than candidates with "black sounding" names. These results held for jobs of all types, and for candidates who were extremely qualified and those who were less qualified. It's no wonder many parents spend so much time choosing their children's names!

---

1   "Are Emily and Brendan More Employable Than Lakisha and Jamal? A Field Experiment on Labor Market Discrimination," by Dr. Marianne Bertrand and Dr. Sendhil Mullainathan, NBER Working Paper No. 9873, July, 2003.

Suggested Answers to Questions

For questions **1** through **16**, answers will vary. Note that on question 12, the best choice of graph—dotplot, stemplot, histogram, or boxplot—will depend on the number of values and the spread of those values. A dotplot or stemplot will display all of the actual data values, but they may be awkward to construct if there are too many data values, or if the values are too spread out. In such cases, a histogram or a boxplot would be easier to construct.

17. This study is not an experiment, because nothing was deliberately done to the students in your class to measure their responses. Instead, students' names, genders, and birth years were recorded, and the corresponding decade ranks data were accessed online.

18. No. The data were collected from students in our class. Since random selection was not used to select the students who took part in the study, we can't be sure that our class represents the larger school population well. We should therefore be hesitant to generalize the results about our class' names to the population of students in our school.

19. If the goal was to generalize to all students at our school, we should use random selection to choose a sample of students from the school population, and then record the names of those students. This could be accomplished using a variation of the hat method with an alphabetical roster of students enrolled at our school. In order to generalize to all high-school students in the district, we would need to use random selection to choose a sample of students from the population of high-school students in the district, and then record the names of the selected students. It might be difficult to obtain a complete list of all high-school students in the district. If so, we could use random selection to choose, say, five students from each high school, and then record the names of those students. To generalize to all high-school students in our state, we would need to use random selection to choose a sample of students from the population of all high-school students in the state. It probably wouldn't be practical to compile a list of all high-school students in the state, or to use random selection to choose a few students from every high school in the state. Instead, we could use random selection to choose, say, 10 high schools from our state, and then use random selection to choose five students from each of those schools.

Possible Extensions

Students might enjoy reading an excerpt from Chapter 6 in *Freakonomics*, by Steven D. Levitt and Stephen J. Dubner, William Morrow/HarperCollins publishers, 2005. The chapter's title is catchy: "Perfect Parenting, Part II; or: Would a Roshanda by Any Other Name Smell as Sweet?"

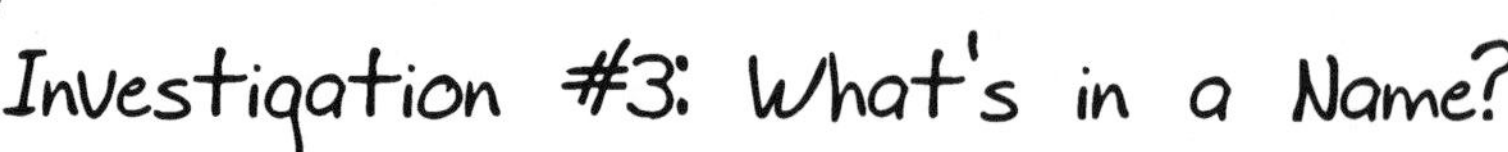 

Investigation #3: What's in a Name?

According to the *Seattle Times* (Oct. 5, 2003), there will be a lot of Jacobs and Emilys in the high-school graduating class of 2020—those were the most popular baby names in the United States in 2002 according to Social Security card applications.

It's nice to be popular, and great to be "cool." The authors of the book *Cool Names for Babies* (Satran, Pamela & Rosenkrantz, Linda, Harper Collins Publishers, 2004) say that it is the unusual names that are most cool.

In this activity, you will carry out an observational study to assess the popularity and coolness of your class based on the names of the students in class.

Getting Started

To complete this activity, you will need to use the Social Security Administration's Popular Baby Name web site. It can be found at *www.ssa.gov/OACT/babynames*.

On this site, you will be able to find lists of the 10 most popular baby names for boys and girls in each year starting in 1880. These lists were compiled using a random sample consisting of 1% of all babies born in a particular year who subsequently applied for a social security card. You will also find a list of the top 1,000 names for each decade from the 1900s to the 2000s.

Spend a few minutes familiarizing yourself with the information available on this web site. Then, start answering the questions that follow.

1. Let's start with an easy question! What is your first name?

2. Are you male or female?

3. In what year were you born?

4. Is your name one of the 10 most popular names for the year in which you were born?

5. Is your name one of the 10 most popular names for the most recent year for which data are available?

Corresponds to pp. 29-32 in Student Module

6. Is your name one of the most popular 1,000 for the decade in which you were born? If so, record your name's rank. If your name is not in the top 1,000, just record that your name is "cool"!

7. After each student in your class has answered questions 1–6, enter the data from the entire class into the following table.

First Name	Gender	Year Born	In Top 10 for Year Born? (Yes or No)	In Top 10 for Most Recent Year? (Yes or No)	Rank for Decade Born (1-1,000 or cool)

8. Is there a most common name for the class? If so, what is the most common name?

9. What is the most common year of birth for the class?

10. In the year that was the most common birth year for the class, what is the most popular name for boys according to the popular baby names web site? For girls? Does anyone in the class have these most popular names?

11. What proportion of the class has "cool" names?

12. Omitting the cool names from the data set, construct a graphical display that shows the distribution of the decade ranks data. How would you describe this distribution? (Comment on shape, center, spread, and any unusual values.)

13. What proportion of the class has names that were in the top 10 names for the year in which they were born?

14. Based on your answers to questions 11 and 13, is your class more popular or more "cool?"

15. What proportion of the class has names that are listed in the top 10 for the most recent year for which data are available?

16. Is the proportion from question 15 lower than, about the same as, or higher than the proportion from question 13? How does this suggest that the popularity of the class' names has changed over time?

17. What makes this study an observational study, rather than an experiment?

18. Was there random selection in the data collection for this study? How does this affect your ability to generalize from the study?

19. How might you modify this study if your goal was to generalize to all students at your school? To all high-school students in your school district? To all high-school students in your state?

Teacher Notes for Investigation #4: If the Shoe Fits ...

THIS INVESTIGATION PUTS STUDENTS IN THE ROLE OF DATA DETECTIVES AS THEY attempt to use a footprint left behind at the scene of the crime to help school administrators identify the perpetrator. First, students examine shoe print length data for a random sample of male and female students. Next, students explore the relationship between height and shoe print length for both male and female students. With their preliminary analysis complete, students must then decide whether the shoe print left at the scene belonged to a male or a female, and predict the height of the culprit. Unlike the two previous investigations, the random selection of students in this observational study allows your students to generalize their findings to the population of all students at the high school.

Prerequisites

Students should be able to:

Describe the shape, center, and spread of the distribution of a quantitative variable using a dotplot and numerical summaries (mean, median; range, inter-quartile range (IQR), standard deviation)

Identify potential outliers

Distinguish an observational study from an experiment

Describe the relationship between two quantitative variables using a scatterplot

Use a scatterplot to make predictions of y from x

Learning Objectives

As a result of completing this investigation, students should be able to:

Explain why random selection in an observational study allows sample results to be generalized to a larger population of interest

Examine the effect of adding a categorical variable to a scatterplot on the relationship between two quantitative variables

Draw conclusions about a population based on graphical and numerical information from a random sample

Make decisions in the face of uncertainty using graphical and numerical information

Teaching Tips

CSI stands for Crime Scene Investigation.

Note that the shoe print lengths were measured in centimeters, but the heights were measured in inches. This was done deliberately to force students to think carefully about units of measurement.

Be sure to emphasize an essential difference between this investigation and the previous two investigations: the use of random selection to produce the data. Random selection is our best attempt to ensure that the sample we choose is representative of the population of interest. Random selection allows us to generalize the results of a sample to the population at large.

Include discussion of using a line of best fit (Q-Q line, median-median line, least-squares line) to make height predictions if your students have the necessary background.

Suggested Answers to Questions

1. Here are comparative dotplots and comparative boxplots of the shoe print lengths from Fathom software.

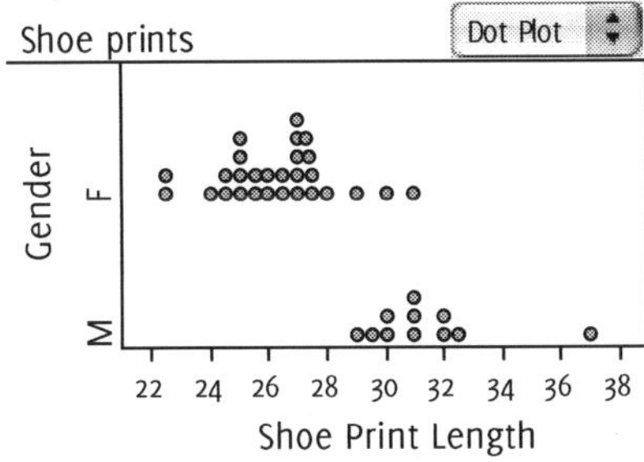

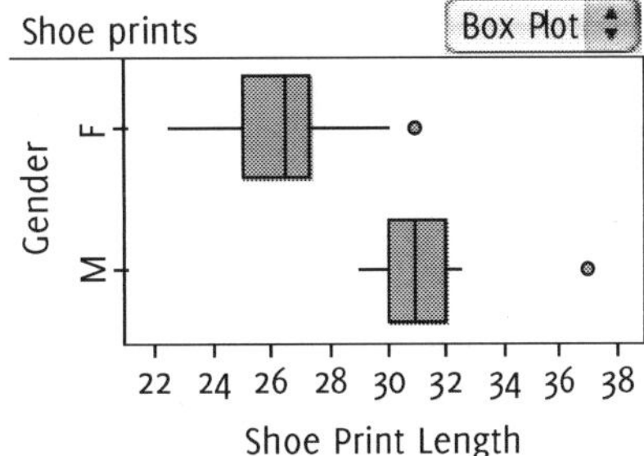

2. We can see from the plots that the males in this sample tended to have longer shoe prints than did the females. A "typical" male in the sample had a shoe print length of about 31 cm, while a "typical" female in the sample had a shoe print length of about 27 cm. There was one male in the sample with an unusually long shoe print—37 cm, or about 14.5 inches! This individual is an outlier according to the $1.5IQR$ rule:

$$Q_3 + 1.5IQR = 32 + 1.5(2) = 35 \text{ is the upper cutoff.}$$

You might be surprised to discover that the female with the longest shoe print is also identified as an outlier by the $1.5IQR$ rule:

$$Q_3 + 1.5IQR = 27.325 + 1.5(2.325) = 30.8125 \text{ is the upper cutoff.}$$

There is slightly more variability in the female shoe print length distribution (range = 8.5 cm; IQR = 2.325 cm) than in the male shoe print length distribution (range = 8 cm; IQR = 2 cm). The shape of the male shoe print length distribution is roughly symmetric with one extremely high outlier. The female shoe print length distribution appears somewhat bimodal.

3. Since nothing was deliberately done to the individuals in this study to measure their responses, this was not an experiment. Instead, the individuals were simply observed and their shoe print lengths, heights, and genders were recorded. That makes this an observational study.

4. The administrators probably wanted to use information from this sample of students to draw conclusions about the population of all students at the school. Their best method for attempting to obtain a representative sample was to choose the students for this study using random selection.

5. A male. There were no females in the random sample with a shoe print length as long as 32 cm, but there were four males with shoe print lengths of 32 cm or more. Although it is possible that a 32 cm shoe print length could be from a female student at the school (not one of the ones in this sample), a more plausible explanation is that the shoe print came from a male student.

6. If the suspect's shoe print length were 27 cm, we would suspect that the suspect was a female. None of the males in the random sample had shoe print lengths less than 29 cm, while about half of the females in the sample had shoe print lengths of 27 cm or less. Although it is possible that a 27 cm shoe print length could be from a male student at the school (not one of the ones in this sample), a more plausible explanation is that the shoe print came from a female student.

If the suspect's shoe print length were 29 cm, we would have a much more difficult time deciding whether the culprit was a male or a female student. A shoe print length of 29 cm falls toward the top end of the distribution of shoe print lengths for the females in the random sample. Three of the 28 females in the sample had shoe print lengths of 29 cm or higher. However, one of the 11 male students in the random sample had a shoe print length of 29 cm. Based on the sample data, it is plausible that a 29 cm shoe print could have come from either a male or female student.

7. The Fathom screen shot below shows the height versus shoe print length data for the students in this random sample. There appears to be a moderately strong linear relationship between shoe print length and height for these students.

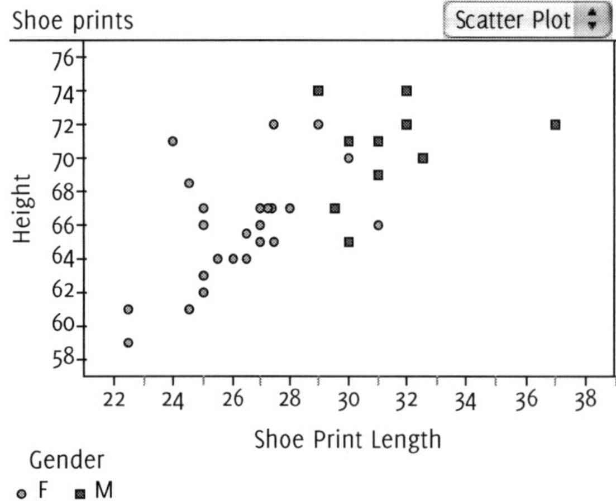

8. No. It appears that two different lines would be needed to summarize the relationship between shoe print length and height—one for females with a steeper slope and lower *y*-intercept, and one for males with a less steep slope but higher *y*-intercept.

9. As students don't know whether the intruder was male or female, they will have to base their predictions on the entire set of sample data. For a shoe print length of 30 cm, the scatterplot suggests a height of about 69 inches. This is close to the average height of the three students in the random sample that had shoe print lengths of 30 cm. In addition, the least-squares regression line for the entire sample of values is:

$$\text{Height} = 44 + 0.832(\text{Shoe print length}).$$

If we substitute 30 for the shoe print length, we get a value of 68.96 inches.

10. From the original doplot, we see that a shoe print length of 31 cm was at the center of the distribution for males in the random sample, but equal to the maximum shoe print length for the 28 females in the sample. As a result, we would infer that the suspect was probably a male. Using the scatterplot of height versus shoe print length, for a male with a 31 cm long shoe print, we would predict a height of around 70 cm. The least-squares line using only the male data is:

$$\text{Height} = 59 + 0.362(\text{Shoe print length}).$$

If we substitute 31 for the shoe print length, we get a value of 70.22 inches. Alternatively, we could have used the average height for the three males in the random sample who had 31 cm long shoe prints: 69.67 inches.

Possible Extensions

1. You might want to have students collect and use data from a random sample of students at your school to draw a conclusion about the perpetrator of the crime. To make shoe prints, have each student step in water and then step on a piece of newspaper. Be sure to measure the length of the shoe print at its longest point for consistency.

2. As a variation on this investigation, you could create a scenario in which a hand print was found at the scene of the "crime," instead of a foot print.

Investigation #4: If the Shoe Fits ...

Welcome to CSI at School. Over the weekend, a student entered the school grounds without permission. Even though it appears the culprit was just looking for a quiet place to study undisturbed by friends, school administrators are anxious to identify the offender and have asked for your help. The only available evidence is a suspicious footprint outside the library door.

Corresponds to pp. 33-36 in Student Module

In this activity, you will use data on shoe print length, height, and gender to help develop a tentative description of the person who entered the school.

After the incident, school administrators arranged for the data in the table below to be obtained from a random sample of this high school's students. The table shows the shoe print length (in cm), height (in inches), and gender for each individual in the sample.

Shoe Print Length	Height	Gender	Shoe Print Length	Height	Gender
24	71	F	24.5	68.5	F
32	74	M	22.5	59	F
27	65	F	29	74	M
26	64	F	24.5	61	F
25.5	64	F	25	66	F
30	65	M	37	72	M
31	71	M	27	67	F
29.5	67	M	32.5	70	M
29	72	F	27	66	F
25	63	F	27.5	65	F
27.5	72	F	25	62	F
25.5	64	F	31	69	M
27	67	F	32	72	M
31	69	M	27.4	67	F
26	64	F	30	71	M
27	67	F	25	67	F
28	67	F	26.5	65.5	F
26.5	64	F	27.25	67	F
22.5	61	F	30	70	F
			31	66	F

Use the data provided to answer the questions that follow.

1. Construct an appropriate graph for comparing the shoe print lengths for males and females.

2. Describe the similarities and differences in the shoe print length distributions for the males and females in this sample.

3. Explain why this study was an observational study and not an experiment.

4. Why do you think the school's administrators chose to collect data on a random sample of students from the school? What benefit might a random sample offer?

5. If the length of a student's shoe print was 32 cm, would you think the print was made by a male or a female? How sure are you that you are correct? Explain your reasoning.

6. How would you answer question 5 if the suspect's shoe print length was 27 cm? 29 cm?

7. Construct a scatterplot of height versus shoe print length using different colors or different plotting symbols to represent the data for males and females. Does it look like there is a linear relationship between height and shoe print length?

8. Does it look like the same straight line could be used to summarize the relationship between shoe print length and height for both males and females? Explain.

9. Based on the scatterplot, if a student's shoe print length was 30 cm, approximately what height would you predict for the person who made the shoe print? Explain how you arrived at your prediction.

10. The shoe print found outside the library actually had a length of 31 cm. Based on the given data and the analysis of questions 1–9, write a description of the person who you think may have left the print. Explain the reasoning that led to your description and give some indication of how confident you are that your description is correct.

Teacher Notes for Investigation #5: Buckle Up

THIS INVESTIGATION ASKS STUDENTS TO EXAMINE THE CHANGE IN SEAT BELT USE IN the states from 2004 to 2005. Data were collected by observers at a random sample of road locations in each state. As in the previous investigation, the random selection should allow students to generalize the sample results to the population of road locations in each state. Students are first asked to consider whether the seat belt use data from the two years should be compared as two separate lists of values, or whether they should analyze the difference in seat belt use from 2004 to 2005 for the states. Then, they must use graphs and numerical summaries to describe the overall change in seat belt use. Students are later asked about the position of their state within the distribution of change in seat belt use. After reviewing details of how the data were produced, students must produce their own summary analysis of the change in seat belt use by drivers.

Prerequisites

Students should be able to:

Describe the shape, center, and spread of the distribution of a quantitative variable using a dotplot and numerical summaries (mean, median, mode; range, interquartile range (IQR), standard deviation)

Identify potential outliers

Distinguish an observational study from a survey or an experiment

Describe the relationship between two quantitative variables using a scatterplot

Learning Objectives

As a result of completing this investigation, students should be able to:

Explain why random selection in an observational study allows sample results to be generalized to a larger population of interest

Decide whether to use a comparative dotplot or a dotplot of differences to analyze data from two samples

Describe the relative standing of one value within a distribution

Explain why an observational study might be preferable to a survey in describing individuals' behavior

Draw conclusions about a population based on graphical and numerical information from a random sample

Teaching Tips

At the time we were writing, the NHTSA's 2007 seat belt use survey was completed, but the data had not yet been released for the individual states. You may want to use more current data from the NHTSA's web site, *www.nhtsa.dot.gov*, for this investigation.

In NHTSA's seat belt usage study, a random sample of road locations within each state was chosen, and then drivers' seat belt use was recorded at those locations. The same method of producing data was used in both 2004 and 2005. Random selection of road locations within each state should allow the seat belt use results to be generalized to all road locations in the state. One value—the overall percent of drivers in the state who were observed wearing seat belts—was recorded for each state in 2004 and again in 2005. These two lists of 48 values are not two independent sets of values. We would expect similar seat belt use within a state from one year to the next. Consequently, we should analyze the differences in seat belt use from 2004 to 2005, and not the two sets of values separately. Through this investigation, we're trying to help students learn an important statistical lesson: The way in which the data were produced determines the proper way of analyzing the data.

Suggested Answers to Questions

1. The Fathom screen shots below show comparative dotplots and comparative boxplots of seat belt use in 2004 and 2005.

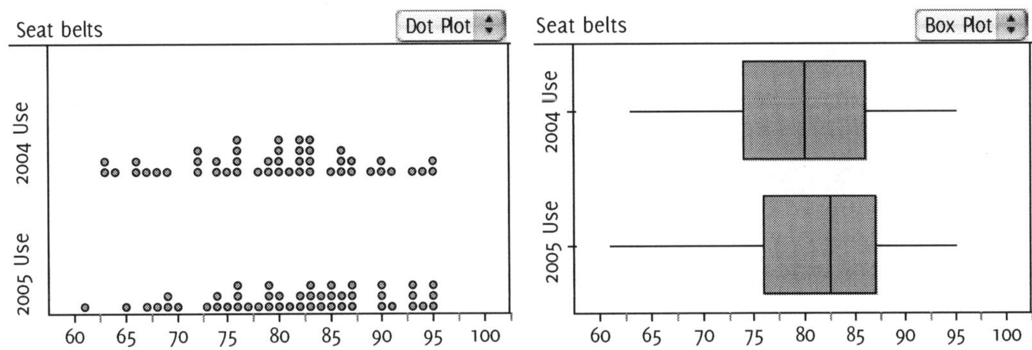

In 2004, between 63% and 95% of the samples of drivers observed in each state were wearing their seat belts. A typical (median) value for seat belt use was 80%. The distribution is roughly symmetric. There are no potential outliers according to the 1.5IQR rule.

In 2005, between 61% and 95% of the samples of drivers observed in each state were wearing their seat belts. A typical (median) value for seat belt use was 82.5%. The distribution is skewed to the left. There are no potential outliers according to the 1.5IQR rule.

The center of the 2005 seat belt use distribution (median = 82.5%) is slightly larger than for the 2004 distribution (median = 80%). Variability in seat belt use among states for the two years is pretty similar (2005 range = 34%; 2004 range = 32%). There appear to be more states toward the higher end of the seat belt use distribution in 2005 than in 2004.

2. A Fathom dotplot and boxplot of the change in seat belt use are shown below. This distribution is skewed to the right. The middle 50% of states showed increases in seat belt use (based on the samples) of between 0% and 3%. Thirty-three of the 48 states showed increases in seat belt use (based on the samples) from 2004 to 2005. The median increase in seat belt use (based on the samples) was 1.5%.

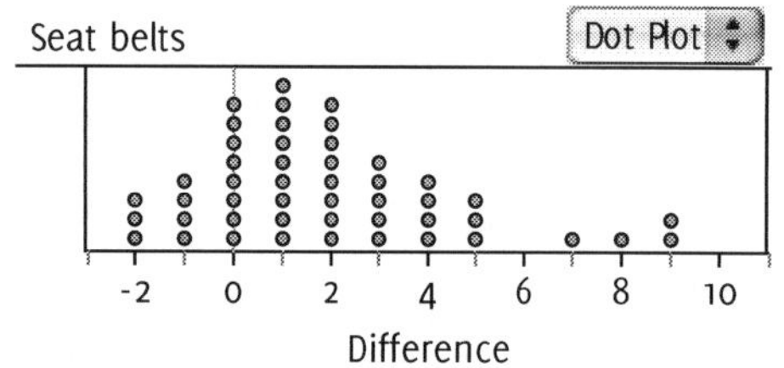

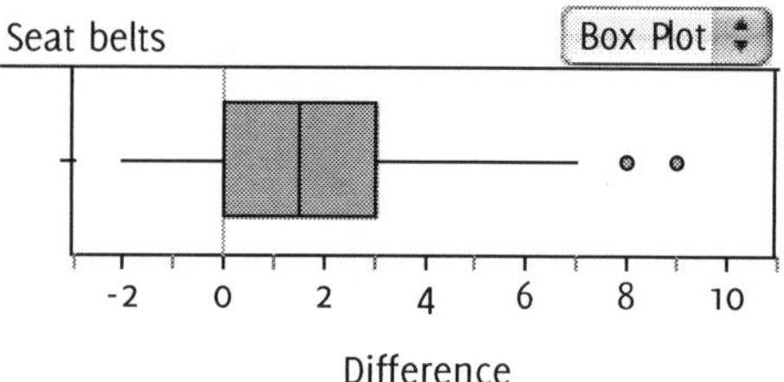

3. The comparative dotplot and boxplot look at the data as two separate lists of values, which hides the paired nature of the data. When we examine the differences in seat belt use via the graphs in question 2, we get a much clearer picture of how seat belt use has changed from 2004 to 2005.

4. Based on the sample results, yes. Thirty-three of the 48 differences are positive, which indicates an increase in seat belt use for those 33 states from 2004 to 2005.

5. Mean = 1.96%; Median = 1.5%

6. The long tail to the right, which includes the three high potential outliers, pulled the mean in that direction. The median is resistant to these extreme values.

7. Use the median, so that the few extremely high values don't make the improvement in seat belt use look better than perhaps it was.

8. There are four potential outliers—7% (Texas), 8% (Nevada), 9% (North Dakota), and 9% (West Virginia). According to the $1.5IQR$ rule:

$$Q_3 + 1.5IQR = 3 + 1.5(3) = 7.5 \text{ is the upper cutoff.}$$

Only Nevada, North Dakota, and West Virginia are identified as outliers. These states showed dramatic increases in seat belt use (based on the samples) from 2004 to 2005.

9. Answers will vary. In New Jersey, for example, seat belt use increased by 4% (based on the samples). This result puts New Jersey in the top 25% of the distribution. So, New Jersey's change in seat belt use is not that typical.

10. Drivers were observed in their vehicles to see if they were wearing seat belts. Drivers were not asked whether they wore seat belts, so this wasn't a survey. Researchers did not deliberately do anything to vehicle drivers to measure their responses, so this wasn't an experiment.

11. Observations of drivers' behavior should produce more accurate information than drivers' self-reported behavior. Some people might claim they wear seat belts when they don't.

12. Because the data were based on observations at a random sample of roadway sites in a given state, we should be able to generalize the results to the population of roadway sites in that state. Note, however, that because the seat belt use value in each state was based on a sample of drivers, the actual seat belt use by drivers in that state might differ a little. A different random sample of roadway sites would probably have led to slightly different seat belt use statistics.

13. Drivers' seat belt use in the states appears to have improved between 2004 and 2005. In that two-year period, 33 of 48 states showed increases in their seat belt use statistics. A typical increase in seat belt use was about 1.5% (the median). The dotplot to the right shows the distribution of changes in seat belt use by state from 2004 to 2005. Note the long right tail, which includes four states with unusually high increases in seat belt use: Texas (7%), Nevada (8%), North Dakota (9%), and West Virginia (9%).

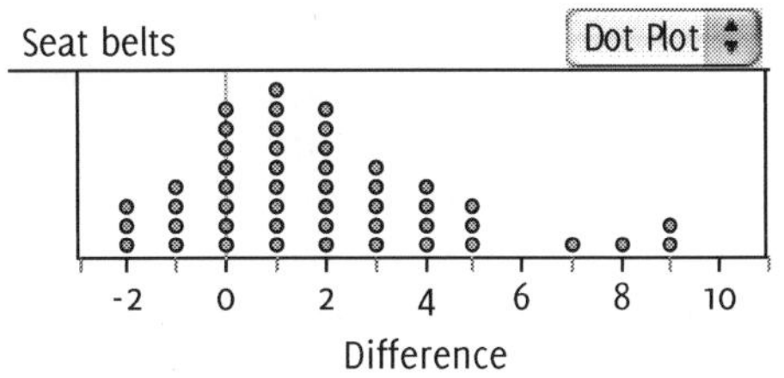

The scatterplot to the right shows a strong, positive, linear relationship between observed seat belt use in 2004 and in 2005. We have added the line $y = x$ for reference. Points above the line represent states with higher observed seat belt use in 2005 than in 2004. Points below the line represent states with lower observed seat belt use in 2005 than in 2004. As most of the points are above the line, we see that observed seat belt use improved in most states during this two-year period.

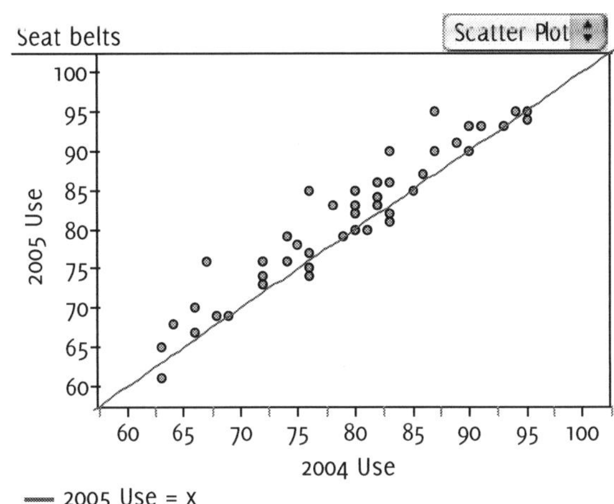

Possible Extensions

1. You might want to have students conduct their own observational study of seat belt use at your school, or in a nearby location. Be sure to get permission from appropriate officials before allowing students to observe drivers' behavior. Also, ensure that students will be safe while making their observations.

2. Students could investigate whether drivers lock their car doors when they park at home versus when they park elsewhere.

Investigation #5: Buckle Up

Corresponds to pp. 37-40 in Student Module

Do you wear your seat belt when driving? Do most people? Is seat belt use changing over time? To answer questions such as these (well, at least the last two questions—only you know the answer to the first question, but we sure hope the answer is yes!), the National Center for Statistics and Analysis published data on seat belt use for 48 states. No data were available for New Hampshire or Wyoming.

The data shown in the table at the top of the next page are from a large-scale study conducted annually by the National Highway Traffic Safety Administration.[1] The study involves actual observation of drivers' seat belt use at a random selection of roadway sites in each state.

The table gives the percentage of drivers observed who used seat belts in 2004 and in 2005. The table also shows the change in seat belt use percentage from 2004 to 2005 (computed as 2005 use percentage – 2004 use percentage).

Use the data in the table to answer the following questions.

1. Would comparative dotplots or comparative boxplots be better for comparing the seat belt use rates for 2004 and 2005? Make the graph that you pick. Then write a sentence or two describing the similarities and differences in the seat belt use rate distributions in 2004 and 2005.

2. Construct an appropriate graph that shows the change in seat belt use by state from 2004 to 2005. Comment on any interesting features of the distribution.

---

1   "Seat Belt Use in 2006—Use Rates in the States and Territories," Traffic Safety Facts, National Highway Traffic Safety Administration, January 2007.

State	2004 Use	2005 Use	Difference	State	2004 Use	2005 Use	Difference
Alabama	80	82	2	Missouri	76	77	1
Alaska	78	83	5	Montana	81	80	-1
Arizona	95	94	-1	Nebraska	79	79	0
Arkansas	64	68	4	Nevada	87	95	8
California	90	93	3	New Jersey	82	86	4
Colorado	79	79	0	New Mexico	90	90	0
Connecticut	83	82	-1	New York	85	85	0
Delaware	82	84	2	No. Carolina	86	87	1
Florida	76	74	-2	North Dakota	67	76	9
Georgia	87	90	3	Ohio	74	79	5
Hawaii	95	95	0	Oklahoma	80	83	3
Idaho	74	76	2	Oregon	93	93	0
Illinois	83	86	3	Pennsylvania	82	83	1
Indiana	83	81	-2	Rhode Island	76	75	-1
Iowa	86	87	1	So. Carolina	66	70	4
Kansas	68	69	1	South Dakota	69	69	0
Kentucky	66	67	1	Tennessee	72	74	2
Louisiana	75	78	3	Texas	83	90	7
Maine	72	76	4	Utah	86	87	1
Maryland	89	91	2	Vermont	80	85	5
Massachusetts	63	65	2	Virginia	80	80	0
Michigan	91	93	2	Washington	94	95	1
Minnesota	82	84	2	West Virginia	76	85	9
Mississippi	63	61	-2	Wisconsin	72	73	1

3. In what way is the graph in question 2 more informative than the graph in question 1?

4. Did most states increase seat belt use from 2004 to 2005? What aspect of the graph you made in question 2 could be used to justify your answer?

5. Compute the mean and median change in seat belt use.

6. What aspect of the graph you made in question 2 explains the large difference between the mean and the median?

7. Would you recommend using the mean or the median to describe the seat belt use change data? Why?

8. Are there any states that stand out as unusual in this data set? If so, which states and what makes them unusual?

9. How did seat belt use in your state change from 2004 to 2005? Would you describe your state as typical with respect to seat belt use change? Explain. (If your state is one of the two states for which no data are given, choose a neighboring state and answer this question for that state.)

10. What makes this seat belt use study observational, rather than an experiment?

11. Why do you think the study was based on actual observation of drivers, rather than a survey of drivers asking if they use a seat belt when driving?

12. Based on the sampling method used in this study, do you think it would be reasonable to generalize the seat belt use results to drivers at all locations in a given state? Explain.

13. Write a brief summary report describing how seat belt use changed from 2004 to 2005. Include graphs and numerical summaries as appropriate.

Teacher Notes for Investigation #6: It's Golden (and It's Not Silence)

In this investigation, students will design and carry out an observational study to determine whether students prefer golden rectangles to nongolden rectangles. Students must first decide on a sampling plan that will allow them to generalize their sample results to all students at the school. After tweaking the plan for practical considerations, they will implement the plan to collect data. With the data in hand, students will perform a preliminary graphical and numerical analysis. Finally, students are asked to prepare a report that summarizes their findings.

Prerequisites

Students should be able to:

Analyze a categorical variable using bar graphs, counts, and percents

Explain how to obtain a true random sample from a population of interest

Learning Objectives

As a result of completing this investigation, students should be able to:

Design a practical sampling plan that incorporates random selection

Conduct an observational study in a way that should produce reliable data

Consider whether the outcome of an observational study could simply be due to chance, rather than an actual preference among individuals

Draw conclusions about a population based on graphical and numerical information from a representative sample

Teaching Tips

For the three rectangles shown on the student handout, the ratio of the longest side to the shortest side is Rectangle #1: 5.33, Rectangle #2: 2.00, Rectangle # 3: 1.60.

Students will need access to a list of students at your school if they are going to select individual students at random. Otherwise, they will need a list that shows other possible sampling units—homerooms, grade levels, math classes, etc. You may need some lead time to acquire the necessary information.

A major purpose of this investigation is to reinforce the benefit of using random selection in choosing a sample—the ability to generalize to a larger population of interest. If it is feasible for students to carry out this observational study using a true random sample from the population of students at your school, they should do so. If true random sampling isn't possible, any practical adjustments your students propose should incorporate random selection. Students often confuse "haphazard" selection with random selection. Remind students that random selection requires a chance mechanism (such as the hat method) be used to choose the individuals in the sample.

There is no prescribed method for determining how many rectangles to use or what dimensions the nongolden rectangles should have in questions 4 and 5. Likewise, we did not specify the number of students who should take part in the observational study. Too few rectangles will make it difficult to distinguish whether students actually have a preference for the golden rectangle or are simply choosing a rectangle at random. Too many rectangles might result in small numbers of students choosing each of the rectangles, which would again make it difficult to detect a preference for the golden rectangle.

Here's an example to help you think through the decisions discussed in the previous tip. Suppose your students opt to use three rectangles, one of which is golden. Suppose further that students decide to choose 30 individuals at random to participate in the study. If individuals actually have no preference among the rectangles in terms of their "goldenness," we would expect about 10 individuals to choose each of the three rectangles. What if 15 students actually say they prefer the golden rectangle? Is it plausible that students are simply picking rectangles at random, and that, just by chance, 15 picked the golden rectangle? Consider rolling a die to simulate such a chance process. Let outcomes 1 and 2 represent students who pick the golden rectangle and outcomes 3, 4, 5, and 6 represent students who pick one of the other two rectangles. Roll the die 30 times, once for each student in the observational study. How often does such a simulation result in as many as 15 people picking the golden rectangle?

(The theoretical probability of 15 or more students picking the golden rectangle if all are choosing at random is about 0.043, using a binomial distribution with $n = 30$ and $p = 1/3$.)

You might want to show students Jim Loy's Most Pleasing Rectangle Poll web page, *www.jimloy.com/poll/poll.htm*, for interesting results about how the orientation of the rectangles might affect people's preferences.

Suggested Answers to Questions

1. To get a true random sample, you would need to obtain a complete listing of all students in the school, and then use random digits or a random number generator to carry out some variation of the hat method. Note that the question asks about obtaining a random sample, not a census of all students at your school.

2. If there are a large number of students in your school, then it may not be practical to collect data from a true random sample of students. If there are a small number of students in your school, then it may be possible to collect data from a random sample.

3. If a true random sample isn't practical, you can still incorporate random selection into your sampling method. For example, if your school has grade-level meetings once per week, you might be able to take separate random samples from each grade to participate in the study. Or, if your school is organized in mixed-grade homerooms (say by alphabetical order of last names), you might be able to take a random sample of homerooms and let every student in the selected homeroom take part in the study. Whatever method you propose should include random selection and offer a reasonable chance of getting a representative cross-section of students from your school.

4. Here are two golden and two nongolden rectangles. The golden rectangles are shaded.

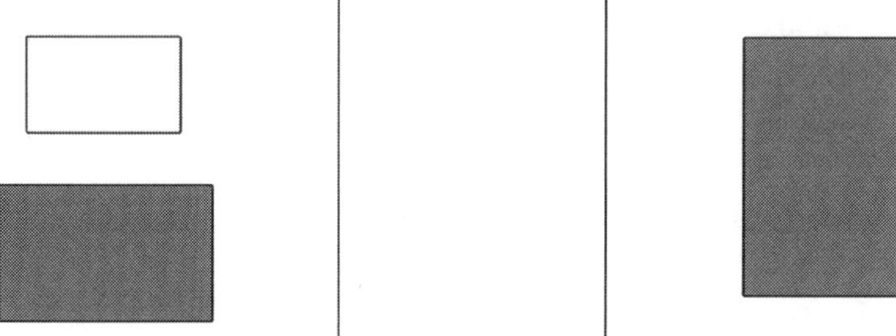

5. Answers will vary. We suggest giving students at least three rectangles from which to choose. If you only provided two rectangles and students really had no preference for one rectangle over the other, they would still pick the golden rectangle half the time, just by chance. This would be equivalent to flipping a coin to choose the preferred rectangle each time. With three rectangles, the chance a student with no particular preference among the rectangles would pick the golden rectangle just by chance is reduced to 1/3. On the other hand, offering the students too many rectangles from which to choose could make it harder to distinguish whether students clearly prefer the golden rectangle.

6. Answers will vary.

7. Answers will vary. Students should summarize individuals' choices in tabular form, showing the number who preferred each of the rectangles, as well as the percent who favored each. Since the variable being measured—preferred rectangle—is categorical, students should present their results graphically in a bar graph.

8. Answers will vary. In evaluating the quality of students' responses, you may want to consider both the accuracy and clarity of communication in:

Tabular presentation of the data

Graphical presentation of the results

Analysis of whether students showed a preference for the golden rectangle

Discussion of the generalizability of results based on their sampling method

Possible Extensions

This observational study can be modified to have students determine whether people are more likely to pick the number "3" when asked to pick one of the numbers 1, 2, 3, or 4.

 # Investigation #6: It's Golden (and It's Not Silence)

Which of the three rectangles shown here do you find the most pleasing?

1

2

3

Corresponds to pp. 41-43
in Student Module

If you picked the third one, you selected the "golden" rectangle. Because they are generally thought to be the most pleasing, golden rectangles are common in art, architecture, and even in the boxes designed for packaging products that are sold in grocery stores.

A rectangle is "golden" if the ratio of its longest side to its shortest side is approximately 1.618.

In this activity, you will design and carry out an observational study to determine if students at your school do, in fact, find golden rectangles more pleasing than other, less-golden ones.

Since the goal is to be able to generalize the study findings to all students at your school, the first thing to think about is how you will select the students who will participate in your study.

1. Describe a way to select study participants that would result in a random sample of students from your school. Don't worry at this point if your plan cannot be easily implemented—instead, focus on what it would take to get a true random sample of students at your school.

2. Do you think it would be possible to actually implement the plan you described in the previous question? Explain.

3. If it would not be possible to carry out the selection plan described in question 1, describe another sampling method that you think would result in a "representative" sample, but not a truly random sample, from your school. Explain why you think a sample selected in the way you propose here could be considered representative of the students at your school.

Now let's think about how you will collect data from the selected students in a way that will enable you to determine if students really do find golden rectangles more pleasing than nongolden rectangles.

4. In the space below, draw a few rectangles that are golden and several nongolden rectangles.

5. In this study, you will be showing the selected students some rectangles and asking which of the rectangles is most pleasing. How many rectangles will you have the selected students choose between? Why did you select this number?

6. Prepare a separate page containing the rectangles to be shown to your study participants.

After your teacher has approved the data collection plan and your page of rectangles, you can proceed to collect the data for your study.

7. Summarize your data in table form and construct an appropriate graphical display of the data.

8. Write a brief report on separate paper that addresses the question "Do students at your school find golden rectangles to be the most pleasing?" Use tables and graphs to support your conclusions.

Teacher Notes for Section II: Surveys

In Section II of the module, students consider a special type of observational study—the survey. As you saw in the previous section, random selection plays an important role here in that it is likely to produce a "representative" sample of some larger population of interest. As with all observational studies, when random selection is used to obtain the sample that will participate in a survey, it is possible to generalize results from the sample to the larger population with confidence. Without random selection, our ability to generalize is limited.

There are three investigations in this section.

Investigation #7: Welcome to Oostburg!

Students use data from a census to answer questions about the corresponding population and then begin to explore the idea of sampling from a population by considering data from a subset of the population.

Investigation #8: Student Participation in Sports

Students draw conclusions from survey data and consider how the way in which a sample is selected affects the conclusions that can be drawn from the resulting survey data.

Investigation #9: Planning and Conducting a Survey—A Class Project

In this culminating investigation, students design, implement, analyze data from, and draw conclusions from a survey to investigate the tooth-brushing behavior of students at their school.

Prerequisites

Students should be able to:

Use census data to compute simple probabilities

Apply proportional reasoning

Compute simple percentages

Construct and interpret a bar graph

Construct and interpret a dotplot

Describe the center, shape, and spread of the distribution of a numerical variable

Compute numerical summary measures, such as the mean and median

Learning Objectives

As a result of completing this section, students should be able to:

Distinguish between a census and a sample

Use survey data to evaluate claims about a population

Explain the importance of random selection when choosing a sample

Explain when it is reasonable to generalize from a sample to the population from which it was selected

Extract information from a comparative bar graph

Explain why it is sometimes best to make group comparisons based on percentages, rather than counts or frequencies

Write simple, clear, and unambiguous survey questions

Develop a reasonable method for obtaining a representative sample of students from their school

Develop and implement a sampling plan

Summarize survey data graphically using bar charts and dotplots

Summarize survey data numerically using measures such as counts and percents for categorical data, or the mean and median for numerical data

Teaching Tips

Throughout the investigations of this section, be sure to emphasize the dangers associated with convenience sampling.

Stress the role of random selection in good sample selection plans.

Encourage students to articulate why they think a particular sampling plan that is being proposed would be likely to result in a sample that could be considered representative of the population.

Obtain administrative permission before allowing your students to conduct the survey in Investigation #9.

Remind students about the ethics of data collection using surveys:

Every individual has the right to refuse to participate in a survey

Preserve the anonymity and confidentiality of individual responses

Possible Extensions

You might want to introduce students to other methods of sampling that involve random selection, such as stratified sampling, cluster sampling, systematic sampling, and multi-stage sampling.

You might also want to conduct a class discussion on the dangers associated with surveys that rely on volunteers or those who self-select to participate in the survey. For example, you might have the class think about the possible problems in generalizing the results of a survey about Internet use that is conducted online or a survey to assess opinion on a controversial topic where people must make a telephone call to participate in the survey.

Section II: Surveys

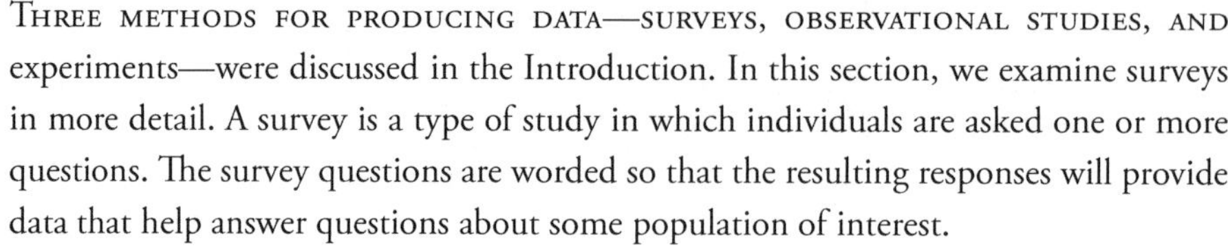

THREE METHODS FOR PRODUCING DATA—SURVEYS, OBSERVATIONAL STUDIES, AND experiments—were discussed in the Introduction. In this section, we examine surveys in more detail. A survey is a type of study in which individuals are asked one or more questions. The survey questions are worded so that the resulting responses will provide data that help answer questions about some population of interest.

If every individual in the population provides responses to the survey questions, the study is called a **census**. A census is the usual method of collecting data only if the population of interest is very small—the students in your math class, for example. However, if the population is large, it is more common for only a subset of the population to provide responses to a survey. In this case, the group of individuals who respond to the survey is referred to as a **sample**.

When only a sample participates in a survey, the way in which the individuals in the sample are selected is critical. As with observational studies, if we want to generalize the results of a survey to the entire population, we need to select the sample in a way that is likely to result in a representative sample.

A popular classic movie called "Magic Town" (1936) featured an actor named Jimmy Stewart playing a very successful pollster. He was able to accurately determine the opinions of the entire United States simply by surveying all the residents of a small town called Magic Town. Because this town was a flawless mirror of the entire country, its residents constituted the perfect sample. Unfortunately for those planning surveys, Magic Town is fictional and much more care needs to go into sample selection!

Just as with observational studies, sample selection can be random or nonrandom. To be reasonably confident that the selected sample will be representative of the population, some type of random selection is required. It is sometimes tempting to select the sample in a nonrandom way just because it is convenient to do so. For example, it might be easy to use the students in your math class as a sample of the students at your high school, but there are many reasons why this sample may not be representative of the entire school—the class may consist of mostly seniors, for example. Because there is no way to tell by just looking at a sample if it is representative of the population, our only assurance comes from the method that was used to choose the sample and from the role that random selection played in the choice.

In addition to being thoughtful about how the sample will be selected, it is also important to think carefully about how the actual survey questions will be worded. Each question should be evaluated to determine if it uses appropriate vocabulary and simple sentence structure and to make sure that the question is clear. This will help to ensure that the survey responses, in addition to being representative of the population, are unambiguous and can be generalized in a straightforward manner.

76

There is one last thing to think about when planning a survey—how large should your sample be? You want the sample to be large enough so that it can reasonably represent the population of interest. On the other hand, it can be both costly and time-consuming to carry out a survey with a large sample size. Because larger samples tend to provide more information than smaller samples, you will need to consider both the desire for a large sample and the available resources for carrying out the survey to arrive at a reasonable sample size.

Planning and carrying out a good survey is a complex task. This overview and the following investigations just provide the basics. You can learn more about surveys in a course in statistics and data analysis. In the investigations that follow, you will explore aspects of planning surveys and analyzing the data that result from them.

In Investigation #9, you will have the opportunity to design and carry out a survey. Collecting survey data involves asking people to share personal opinions or ideas. Not everyone feels comfortable doing that. Any individual has the right to refuse to participate in a survey. When you are in the role of researcher, you must respect that right. It is also your responsibility to preserve the anonymity and confidentiality of responses.

Teacher Notes for Investigation #7: Welcome to Oostburg!

THIS INVESTIGATION INTRODUCES STUDENTS TO SURVEY DATA RESULTING FROM A CENSUS of a small town. Students use the data to compute various probabilities and to evaluate the accuracy of claims that have been made about the town. In the second part of the activity, students are given survey data resulting from a sample from the population and are asked to compare results based on this sample to population results. Finally, students are led through a discussion of why convenience samples are not a good way to collect data if the goal is to generalize to the larger population.

Prerequisites

Students should be able to:

Use census data to compute simple probabilities

Apply proportional reasoning

Learning Objectives

As a result of completing this investigation, students should be able to:

Distinguish between a census and a sample

Use survey data to evaluate claims about a population

Explain the importance of random selection when choosing a sample

Explain when it is reasonable to generalize from a sample to the population from which it was selected

Teaching Tips

When discussing question 2, you may need to help students see how they can use proportional reasoning to compute the desired estimate. You can give an example like the following: There are 38 + 46 = 84 people in the 41 to 60 age group. If we were interested in how many were between 41 and 45 years old, a reasonable estimate would be ¼(84) = 21, because 41 to 45 is about ¼ of the 41 to 60 age range.

Question 3 is a good place to remind students that they can use what they know about sets (in particular about the complement of a set) to help them compute the desired probability.

Question 3 is also a good place to remind students that there is a difference between the probability that a randomly selected person is male and did not watch "The Simpsons" and the probability that a randomly selected male did not watch "The Simpsons." The latter probability is a conditional probability (the probability of not watching "The Simpsons" *given* that the selected person is male), and the two probabilities are not the same.

Encourage group discussion in the evaluation of the headlines in question 5. You might begin by asking how many students think a particular headline is accurate and how many think it is not accurate. Then choose one student on each side to argue their case. A consensus should then develop about each headline.

Be sure to have a class discussion about the dangers associated with convenience sampling. Ask students to think of a good way and a poor way of selecting a sample of students at your school if they plan to use the resulting sample to estimate how much time students spend in the school library.

Suggested Answers to Questions

1. (a) The total number who attended a movie is $4 + 5 + 16 + 22 + 12 + 22 + 4 + 7 = 92$. Because the data represent a census of the 306 people in the town,

$$P(\text{selected person attended a movie}) = 92/306 = .301$$

(b) The number who attended a movie and who are also between 18 and 40 years old is $16 + 22 = 38$, so

$$P(\text{selected person attended a movie and is between 18 and 40 years old}) = 38/306 = .124$$

(c) The total number of males is $36 + 32 + 38 + 32 = 138$, so

$$P(\text{male}) = 138/306 = .451$$

(d) The total number who are between 18 and 60 years old is $32 + 35 + 38 + 46 = 151$, so

$$P(\text{selected person is between 18 and 60}) = 151/306 = .493$$

2. You might first want to address why this estimate is more difficult. The age range 10 to 30 does not match the age groups used to summarize the survey data. As a result, we do not know the exact number of people in this range and can only estimate. A student may indicate that any guess is possible, but a more acceptable answer (one used in applications involving census data) is to determine a reasonable estimate of the number of people in the age range based on the totals for the reported categories. For this question, the 10- to 17-year-old age group is very close to half the ages represented in the category "17 and younger," so an estimate of one-half of the 77 people in the 0–17 age range would be a reasonable estimate (or, approximately 35 to 40 people). Similarly, 18 to 30 years is approximately half the number of ages in the range of 18 to 40. (Some students might be more precise with this proportion.) Therefore, an approximation based on half of the 67 people in that age range would be reasonable (or, 30 to 35 people). If we put these estimates together, we get an estimate of 65 to 75 people in that age range. This would indicate a probability of approximately .19 to .23. As all answers are estimates, evaluate students' thinking based on their use of proportional reasoning.

3. There are a total of 138 males in the town. Of these, $23 + 29 + 12 + 1 = 65$ reported watching "The Simpsons." So, $138 - 65 = 73$ males did not watch "The Simpsons."

$$P(\text{selected person is male and did not watch "The Simpsons"}) = 73/306 = .239$$

4. Answers will vary depending on questions posed.

5. (a) This is an accurate headline as 162 of the eligible voters turned out for the last election. Since there are 229 eligible voters, 162/229, or approximately .707 or 71% of the eligible voters, indicated they voted.

(b) This is also an accurate headline, as one out of 78 people in the over 60 age category indicated that they shopped for clothes online (or approximately 1%), compared to 54 out of 67 people (or approximately 81%) in the "18–40 years old" category.

(c) Most interpretations would indicate that this is not accurate based on a comparison of the proportion who attended a movie for each age group, or

0–17 years old: 9/77 or ≈ 11.7%

18–40 years old: 38/67 or ≈ 58%

41–60 years old: 34/84 or ≈ 41%

Over 60 years old: 11/78 or ≈ 14.1%

The variation in percents is rather large, indicating that the generations represented by these age categories are not attending movies in the same way.

However, if the headline is interpreted as simply stating that people from each generation attend movies, then yes, the headline is accurate. Ask students how they would interpret this headline.

(d) Most interpretations would indicate that this headline is accurate. Comparing the "18 to 40 years old" group to the "41 to 60 years old" group would be a good indicator to support the headline, or

61/67 (≈ 90%) ate fast food in the "18 to 40 years old" group

13/84 (≈ 16%) ate fast food in the "41 to 60 years old" group

(e) Most interpretations would indicate that this is accurate. Comparing the "over 60 years old" group to the "18–40 years old" group would support this answer:

3/78 (≈ 4%) watched "The Simpsons" in the "over 60 years old" group, compared to 60/67 (≈ 90%) in the "18–40 years old" group.

(f) This is clearly not an accurate headline. 72/78 (≈ 92%) of the people over 60 voted in the last election. This is not close to 40%.

(g) Most interpretations of this question would indicate that it is accurate. This would be based on the variation in the percent of people in each age category who indicated that they eat out:

0–17 years old: 13 out of 77 people (≈ 17%)

18–40 years old: 61 out of 67 people (≈ 90%)

41–60 years old: 13 out of 84 people (≈ 15%)

Over 60 years old: 7 out of 78 people (≈ 9%)

6. Probably not. Oostburg has gender and age distributions that are not similar to the country as a whole.

7. Yes, Hugo was accurate, as 28 people responded yes to the question. Therefore, 28/40 = 70%.

8. No. Only 77/306 (or approximately 25.2%) of the total town indicated that they shopped online. The difference in the sample Hugo collected and the census data indicates Hugo is not accurately summarizing the town. This is very likely due to the method Hugo used to select his sample.

9. As the percent of people who shopped online from the sample is similar to the percent of the age category of 18 to 40 years old in the census data, it is likely the band is made up primarily of people who are 18 to 40 years old. This would make sense if the band in which Hugo is a member is a high-school band (as opposed to a community band, which might have members from all age categories).

10. With random selection, we can be confident that the resulting sample will be representative of the population.

11. Answers will vary. Good answers will employ some form of random selection and will provide a convincing argument that the proposed method will result in a representative sample.

Investigation #7: Welcome to Oostburg!

Oostburg is a small town in Wisconsin. The 306 residents of this town are very data-driven! They are willing and anxious to respond to surveys and give their opinions about various issues. A recent survey was conducted in Oostburg and every person who lives there responded. (Although baby Edna, the youngest citizen of Oostburg at only 8 months old, was not able to answer any of these questions, her parents were willing to respond for her.) This particular survey included questions about age, sex, voting behavior, and participation in various activities during the last month. Data from the survey are summarized in the following two tables.

Corresponds to pp. 46-50 in Student Module

Age	17 and Younger		18 to 40 Years Old		41 to 60 Years Old		61 and Older	
Sex	Male	Female	Male	Female	Male	Female	Male	Female
Number of Responses	36	41	32	35	38	46	32	46

Age	17 and Younger		18 to 40 Years Old		41 to 60 Years Old		61 and Older	
Sex	Male	Female	Male	Female	Male	Female	Male	Female
Voted in last town election	0	0	10	12	28	40	29	43
Attended a movie during the last month	4	5	16	22	12	22	4	7
Ate fast food at least once during the last month	6	7	28	33	8	5	3	4
Shopped for clothes online during the last month	5	6	26	28	4	7	0	1
Watched "The Simpsons" during the last month	23	27	29	31	12	8	1	2

What do the data tell us about Oostburg residents? Given that the entire population of Oostburg was surveyed, the above data is a census of the town. Use the given data to answer the following questions.

1. If a resident of Oostburg is to be selected at random, what is the probability that the person selected:

(a) attended a movie during the last month?

(b) attended a movie and is 18 to 40 years old?

(c) is male?

(d) is between 18 and 60 years old?

2. Estimate the probability that a person selected at random is between 10 and 30 years old. Why is this probability more difficult to compute than those of question 1?

3. What is the probability that a person selected at random is male and did not watch "The Simpsons" in the past month?

4. Pose two other probability questions that could be answered using the survey data and then answer those questions by computing the relevant probabilities.

5. Below are seven headlines from the *Oostburg Herald*, a local newspaper. Evaluate the accuracy of each headline based on the survey data. Write a sentence or two giving your assessment of the headline, using the survey data to support your evaluation.

(a) "70% of Eligible Voters Turned Out for Election" (Assume the eligible age of voting in Oostburg is 18.)

(b) "Over 60 Crowd Not Responding to Online Shopping"

(c) "Movies Are Reaching Across ALL the Generations"

(d) "Fast Food Eating a Big Thing with the Younger Crowd"

(e) "'The Simpsons' Not Popular with Older TV Viewers"

(f) "40% of People Over 60 Voted in the Election!"

(g) "Oostburgians Eating Preferences Dependent on Age!"

6. The section overview describes "Magic Town," a town that is a flawless mirror of the entire country. Do you think Oostburg could be such a magic town? Explain your reasoning.

Hugo VanHorn, a senior at Oostburg High School, did not have access to the data from the survey described here. For a school project, Hugo decided to investigate the popularity of online shopping in Oostburg. After band practice, he quickly asked 40 band members if they had shopped for clothes online in the past month. The results from his survey are summarized below:

Have you shopped for clothes online during the past month?	
Yes	No
28	12

Hugo was quite impressed with his results so he wrote a report about the popularity of online shopping in Oostburg. His report indicated that 70% of the residents of Oostburg had shopped for clothing online in the past month.

7. Is the statement that 70% shopped for clothing online in the past month an accurate summary of Hugo's sample? Explain your answer.

8. Is Hugo's statement that 70% of all Oostburg residents shopped for clothing online during the past month an accurate statement? Justify your answer.

9. Hugo's sample was a convenience sample; he did not randomly select his survey participants from the residents of Oostburg. As a consequence, Hugo's sample was not representative of the Oostburg population. In fact, residents in one of the age groups were over-represented in his sample. Based on the census survey data, which age group do you think was over-represented in Hugo's sample? Explain your reasoning.

10. Why would it have been better for Hugo to have used random selection in choosing the 40 people who would participate in his survey?

11. Assuming that Hugo would like to be able to use survey data to generalize to the Oostburg population, write a brief set of instructions that Hugo could use to select 40 participants for a new survey.

Teacher Notes for Investigation #8: Student Participation in Sports

IN THIS INVESTIGATION, STUDENTS USE SURVEY DATA FROM A RANDOM SAMPLE TO investigate the plausibility of statements about a population. Students are also asked to think about why methods of choosing a sample that do not use random selection may result in an unrepresentative sample. Finally, the idea of sampling variability is introduced, setting the stage for an introduction to inference in Section IV.

Prerequisites

Students should be able to:

Interpret a bar graph

Compute simple percentages

Learning Objectives

As a result of completing this investigation, students should be able to:

Extract information from a comparative bar graph

Use survey data to evaluate claims about a population

Explain the importance of random selection when choosing a sample

Explain why it is sometimes best to make group comparisons based on percentages, rather than counts or frequencies

Teaching Tips

Focus the discussion of question 6 answers on why the proposed methods, which do not involve random selection, may yield a sample that is not representative of the population.

Question 9 is a good place to introduce the idea of sampling variability. Even if 50% of the students at the school are female, we would not expect the percentage of females in a sample to be exactly 50%.

Question 10 begins to hint at the ideas that will be formalized in Section IV. Answering this question requires an informal assessment of whether a sample percentage of 66% could have resulted due to just sampling variability when the population percentage is 50%. Here, appeal to the students' intuition. With random selection and a sample size of 50, a sample percentage of 66% is unlikely to have occurred by chance. Point out that more formal methods for making such assessments will come in Section IV.

In the discussion of questions 11 to 13, make the case for using percentages to compare groups, rather than making the comparison based on raw frequencies.

Suggested Answers to Questions

1. 18

2. 15

3. 12

4. 5

5. Answers will vary.

6. (a) Answers will vary, but should address possible reasons why such a sample might not be representative of the population. For example, maybe some sports teams practice before classes start in the morning, so athletes may be over-represented in a sample that consists of the first 50 students to arrive on campus.

(b) Answers will vary, but should address possible reasons why such a sample might not be representative of the population. For example, maybe the two sections of pre-calculus are offered in the afternoon only, when athletes are typically required to travel to athletic events.

(c) Answers will vary, but should address possible reasons why such a sample might not be representative of the population. For example, it is more likely that students in the weight room would be athletes, making this sample not representative of the general student population.

7. 33/50 = .66 = 66%

8. 17/50 = .34 = 34%

9. (1) There may be more females than males at the school, and (2) the percent of females in a sample will not necessarily be exactly equal to the actual percent of females in the population (sampling variability).

10. No. There were twice as many girls in the sample as there were boys. Because the sample was randomly selected and the sample size was 50, it is not likely that we would see twice as many girls in the sample if the number of boys and the number of girls at the school were about the same.

11. Yes. In the sample, there were 18 girls and only 12 boys who reported participating in sports.

12. The greater number of girls in the sample who participated in sports could just be due to there being more girls in the sample. The proportion of girls in the sample who participated in sports is 18/33 = .55. For boys, this proportion is 12/17 = .71.

13. There were more girls in the sample who reported participating in sports (18 for girls versus 12 for boys). However, the greater number of girls in the sample who participated in sports could just be due to there being more girls in the sample. The proportion of girls in the sample who participated in sports is 18/33 = .55. For boys, this proportion is 12/17 = .71.

14. Answers will vary.

 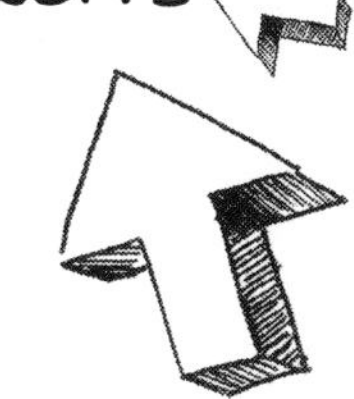

Investigation #8: Student Participation in Sports

The short article below is from the student newspaper at Rufus King High School. Use the information from the article to answer the following questions.

Corresponds to pp. 51-54 in Student Module

Student Survey Finds Females More Involved in Sports

By Kayla Johnson

A Rufus King mathematics class conducted a survey to investigate student participation in extra-curricular activities. Fifty randomly selected students participated in the survey.

One question on the survey asked about participation in school sports programs. The accompanying graph shows participation by sex.

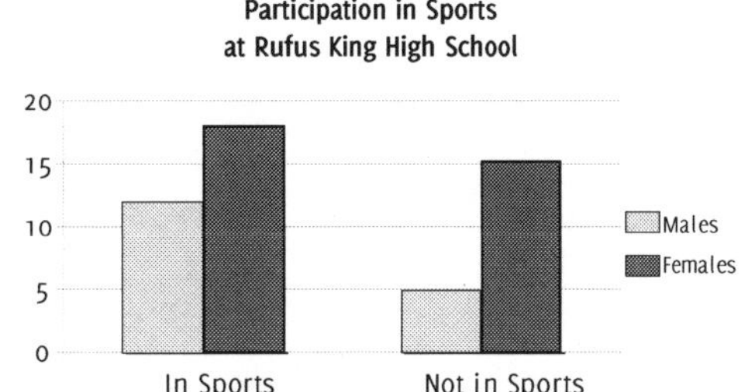

Shauna Rafferty, a junior at Rufus King and captain of the girls' soccer team, noted that there were more females in the sample that participated in sports. She said, "I think the girls in our school are more active in sports than the males. I am sure the success of our soccer team played a large role in this increased interest."

Bryon Jones, a junior on last year's state championship boys' basketball team did not agree. He responded, "All of the guys in my classes belong to one of the sports programs in the school."

Mr. Samuelson, the athletic director of the school, indicated that the number of sports programs and the number of students participating in the programs has posed a real problem in scheduling practice sessions. "Hopefully we will not have to eliminate some of the programs available to our students because students are not able to get adequate practice session time," Mr. Samuelson indicated.

Mr. Samuelson further stated that it was difficult to balance the demand for time in the weight room, the gym, and the outdoor fields.

1. How many of the 50 students surveyed were females involved in sports?

2. How many of the 50 students surveyed were females not involved in sports?

3. How many of the 50 students surveyed were males involved sports?

4. How many of the 50 students surveyed were males not involved in sports?

5. The article says that the 50 students surveyed were randomly selected. Describe one way in which this random selection might have been accomplished.

6. For each of the sample selection methods listed below, give two reasons why random selection of survey participants would be preferable.

(a) Give the survey to the first 50 students who arrive on campus on a Friday morning.

(b) Give the survey to all of the students enrolled in the school's two sections of pre-calculus.

(c) Give the survey to all students who use the weight room on a particular day.

7. What percent of the 50 students in the survey sample were female?

8. What percent of the 50 students in the survey sample were male?

9. What are two reasons that the percent of females in the survey sample might not be 50%?

10. Based on the results of the survey, do you think that the number of girls attending the school is about the same as the number of boys attending the school? Explain your reasoning.

11. Is Shauna correct in her statement that there were more girls in the sample who participated in sports than boys who participated in sports?

12. Explain why Shauna may not be correct in her statement that the survey results imply that girls at the school are more interested in sports than boys.

13. The headline in the school newspaper states more females participate in sports. Explain how this statement could be considered accurate and explain how this statement is at the same time misleading.

14. Write a replacement headline that is not misleading, and then write a few sentences that you think accurately summarize the survey results.

IN THIS CULMINATING INVESTIGATION FOR SECTION II, STUDENTS DEVELOP AND implement a sampling plan to investigate two aspects of tooth-brushing behavior—the reported length of time that students spend brushing and whether students leave water running while brushing.

Prerequisites

Students should be able to:

Construct and interpret a bar graph

Construct and interpret a dotplot

Describe the center, shape, and spread of the distribution of a numerical variable

Compute numerical summary measures, such as counts and proportions for categorical data, or the mean and median for numerical data

Learning Objectives

As a result of completing this investigation, students should be able to:

Write simple, clear, and unambiguous survey questions

Develop a reasonable method for obtaining a representative sample of students from their school

Develop and implement a sampling plan

Summarize survey data graphically using bar charts and dotplots

Summarize survey data numerically using measures such as the mean and median

Teaching Tips

Obtain permission from your administration, if needed, before having your students carry out the survey.

Remind students about the ethics of survey data collection, including an individual's right to "opt out" and the importance of preserving respondents' anonymity and confidentiality.

In the discussion of the survey questions, focus on whether the questions are clear and unambiguous and whether they will elicit the desired information. For example, the question about brushing time should specify that the response should be in seconds.

Sometimes it is difficult to obtain a true random sample from a population. In this investigation, it may be impractical to select at random from a list of students at the school and then somehow find the selected students to participate in the survey. So, other methods may need to be considered. In a similar situation, for example, a student suggested the following plan:

Post students at each entrance to campus and then (1) use a random number generator or table of random numbers to obtain a random number between 1 and 19, (2) use the random number generated in step 1 to determine the next student to be approached (for example, if the random number is 8, the students would let the next 7 students go by and then approach the next student—the 8th student—to participate in the survey), (3) repeat steps 1 and 2, each time generating a new random number.

While not an "official" random sample, this method uses a form of random selection and would be a reasonable alternative.

The class discussion of question 13 is a good place to revisit the idea of sampling variability by talking about how the sample mean will vary somewhat from sample to sample.

Suggested Answers to Questions

1. Answers will vary.

2. Answers will vary.

3. Answers will vary.

4. Answers will vary. It may not be practical to select a true random sample, but some sort of random selection should be employed. For an example, see the Teaching Tips above.

5. Answers will vary.

6. Answers will vary.

7. Answers will vary. Answers should appeal to the way random selection was employed and should argue why the student believes the sampling plan is likely to produce a representative sample.

8. Answers will vary. Even if students leave the water on, they may be tempted to say they turn if off because they know this is the right thing to do from a conservation standpoint.

9. Answers will vary. Answers should comment on how the two proportions (for females and for males) compare.

10. Answers will vary. Answers should comment on center, shape, and spread.

11. Answers will vary. Answers should be based on the dotplot and whether the responses for females tend to fall to the right end of the dotplot or whether they are intermingled with the responses from males.

12. Answers will vary.

13. Answers will vary.

14. Answers will vary. Students will need to think about how to measure actual brushing time. This may be difficult to do, as having an observer present or asking someone to time his or her brushing time may actually influence behavior.

# Investigation #9: Planning and Conducting a Survey

In this investigation, you will develop a sampling plan and carry out a survey to investigate tooth-brushing behavior of students at your school. Consider the following two recommendations.

From *www.animated-teeth.com*:

> As you might guess, many humans simply aren't self-disciplined enough to brush properly when they use a manual toothbrush. As a general rule, most people should brush their teeth at least twice a day with each **brushing period encompassing at least two to three minutes**. The fact of the matter is that most of us fail to routinely meet these guidelines.

From a *Los Angeles Daily News* (December 15, 2007) article titled "Water District Asks Users to Cut Back by 10 Percent. Drought Depleted Supplies Spur Voluntary, Mandatory Measures to Limit Consumption"

> The Las Virgenes Water District is asking residents to reduce water use by 10 percent and is ordering farmers to cut back by a third. In seeking voluntary and mandatory cutbacks, the district follows the lead of Long Beach and other cities responding to an ongoing drought. "With no relief to the drought in sight, we must take steps now to ensure we have adequate supplies for the coming year," said John Mundy, the district's general manager. "We are dealing with water cutbacks throughout the state." … Since nearly 70 percent of water is used outdoors, the district is asking residents to reduce use, water every other day and to sweep, rather than hose off, driveways. They also called upon residents to fix leaks, take shorter showers, and **shut off faucets while shaving or brushing teeth**.

1. Write a set of survey questions that would allow you to get responses regarding the following three characteristics of selected students at your school:

Sex of the survey participant

Whether or not the survey participant leaves the water on or turns the water off while brushing his or her teeth

How long, in seconds, the survey respondent thinks that he or she spends when brushing his or her teeth

You can also include other questions you think might be of interest.

Corresponds to pp. 55-61 in Student Module

2. As a class, discuss the proposed survey questions and come to an agreement on the wording of the questions to be included in the survey. Record the final version of the survey questions below.

3. As a class, discuss whether you think it would be easy or difficult to obtain a random sample of 50 students at your school and to obtain the desired survey information from all the students selected for the sample. Write a few sentences summarizing the class discussion in the space below.

4. As a class, decide how you will go about selecting a sample of 50 students that reasonably could be considered representative of the population of students from your school. Write a brief description of the sampling plan, and point out the aspects of the plan that make it reasonable to argue that it will be representative.

5. Carry out the survey and record the responses in the table below. If you included additional questions in your survey, you can modify the data sheet as needed. As a reminder: *Collecting survey data involves asking people to share personal opinions or ideas. Not everyone feels comfortable doing that. Any individual has the right to refuse to participate in a survey. When you are in the role of researcher, you must respect that right. It is also your responsibility to preserve the anonymity and confidentiality of students' responses.*

Survey Data

Respondent	Sex (M or F)	Water Off (Y or N)	Time Spent Brushing	Respondent	Sex (M or F)	Water Off (Y or N)	Time Spent Brushing
1				26			
2				27			
3				28			
4				29			
5				30			
6				31			
7				32			
8				33			
9				34			
10				35			
11				36			
12				37			
13				38			
14				39			
15				40			
16				41			
17				42			
18				43			
19				44			
20				45			
21				46			
22				47			
23				48			
24				49			
25				50			

Now use the survey data to answer the following questions.

6. Construct a bar chart of the "water off" data. What proportion of survey respondents reported that they turn the water off while brushing their teeth?

7. Think for a minute about how the students in the sample were chosen. Do you think the proportion of students at your school who report that they turn the water off while brushing is likely to be much smaller than, much larger than, or somewhere near the value of the proportion computed in question 6? What aspect of the survey design supports your answer?

8. Sometimes there is a difference between what people say they do and what they *actually* do. Do you think this might be the case for the "water off" question? Explain your reasoning.

9. What proportion of the girls in the survey sample report that they turn the water off while brushing? How does this compare to the proportion of boys that say they turn off the water?

10. Use the reported brushing time data to construct a dotplot. Write a few sentences describing what the dotplot tells you about the distribution of brushing times.

11. Now construct a dotplot that uses color to distinguish between the reported brushing times of females and the reported brushing times of males (use one color for dots that correspond to responses that came from females and a different color for the dots that represent responses from males). Does this plot suggest that females tend to report longer brushing times? Explain.

12. Find the median of the data set consisting of the 50 reported brushing times. Divide the survey responses into two groups—those whose reported brushing times were less than the median brushing time and those whose reported brushing times were equal to or greater than the median brushing time. Use the table below to organize the information needed to compute the proportion that report turning off the water while brushing for each of these two groups. Do these proportions suggest that people who brush longer may be more likely to turn off the water while brushing? Explain.

	Below the Median Brushing Time	**Equal to or Greater than the Median Brushing Time**
Number in the Sample		
Number Who Report They Turn Off Water		
Proportion Who Report They Turn Off Water		

13. The web site referenced earlier (*www.animated-teeth.com*) also included the following:

> Actually, the statement that most people aren't self-disciplined enough to brush properly when they use a manual toothbrush is probably a little bit harsh. Research has found that there can be a major discrepancy between the amount of time that a person actually does brush, as compared to the amount of time that they perceive they have brushed.
>
> One study (*Journal of Clinical Dentistry*, 1998, 9(2):49-51) found that their test subjects, on average, brushed their teeth for 78 seconds (a little longer than a minute) when they actually thought they were brushing for 141 seconds (over two minutes, an adequate amount of time). So, the intention of these people was appropriate but in reality their actions (actual brushing time) were lacking.

Compute the mean of the 50 reported brushing times in the survey data set. How does your sample mean compare to the value of 78 seconds in the quote above?

14. As a class, discuss how you might design a study that would help you determine if there is a discrepancy between reported brushing times and actual brushing times for students at your school. Write a few sentences summarizing the class discussion.

Teacher Notes for Section III: Experiments

Observational studies are a very useful tool for gaining information about the world around us. With observational studies, we can determine the answers to many questions: Has drivers' seat belt use in the states increased since last year? How often do people wash their hands after using the bathroom? Is there a relationship between the height and shoe print length of teenagers? Are meat hot dogs less healthy than poultry hot dogs?

But there are many other important questions that observational studies cannot help us answer. Here are a few: Does smoking cause lung cancer? Is a new medication for treating migraine headaches more effective than the current treatment that doctors most often prescribe? Which is more effective for reducing weight in obese adults, a low-fat diet or a low-carbohydrate diet? Does listening to Mozart help people memorize better than working in silence? To get answers to these questions, which suggest some sort of cause-and-effect conclusions, we must perform experiments. In this section, students will examine experiments in more detail.

There are three investigations in this section.

Investigation # 10: Do Diets Work?

This investigation presents students with the results of two well-designed experiments that compared the effectiveness of low-carb and low-fat diets in reducing weight and lowering cholesterol in obese adults. Students are led in a step-by-step fashion to identify and describe specific design elements of these two experiments. Then, students are asked to interpret the results of the two studies in context, taking into account some possible limitations of each study.

Investigation #11: Distracted Learning

Students begin by incrementally designing an experiment to test whether listening to Mozart improves performance on a memorization task. With their design established, students carry out the experiment using the members of their class as subjects. Once the data have been produced, students set about the task of analyzing and drawing conclusions from the data.

Investigation #12: Would You Drink a Blue Soda?

This investigation serves as a culminating investigation on experiments. From what they have learned in the Overview and from completing investigations #11 and #12, students should be ready to design an experiment on their own. This time, they can use random selection to choose subjects, which will extend their ability to generalize results to a larger population of interest.

Prerequisites

Students should be able to distinguish an observational study from a survey or an experiment.

Learning Objectives

As a result of completing the Overview, students should be able to:

Identify the experimental units/subjects, factor(s)/explanatory variable(s), treatments, and response variable(s) in an experimental setting

Explain the purpose of randomly assigning treatments to subjects in an experiment

Determine whether an experiment was carried out in a single-blind or double-blind manner

Explain the purpose of control in an experimental design

Explain what is meant by "statistical significance"

Explain why replication is an important experimental design principle

Identify a potential confounding variable in a study and explain how the variable could result in confounding

Explain how the way in which data were produced affects our ability to generalize results to a larger population of interest

Teaching Tips

The Overview is chock full of important terminology and issues related to experimental design. Our advice is to have students take turns reading the material aloud, pausing at appropriate spots to clarify definitions and ideas.

In the first paragraph, we distinguish an experiment from an observational study. Simply put, an experiment requires that researchers deliberately impose specific conditions and measure some response.

As our first example of an experiment, we consider a biologist who wants to compare the effects of two brands of weed killer on a particular variety of broad-leafed plant found in a university's garden. A primary purpose of this example is to convince students that the method of assigning treatments to experimental units is vitally important. More specifically, we argue that the "best" method of determining which experimental units receive which treatments is to let chance decide. This process of "random assignment" gives researchers the best hope of starting out with fairly equivalent groups of experimental units prior to administering treatments. Without random assignment, researchers risk creating groups of experimental units that differ in some important way that could systematically affect their response to the treatments. Then, any differences in response between the groups could be due to these initial differences, rather than to the effects of the treatments. This circumstance, in which the effects of the treatments are hopelessly mixed up with the effects of some other variable on individuals' responses, is known as *confounding*. Random assignment of treatments to experimental units gives researchers a powerful tool for avoiding confounding.

Random assignment also helps with the primary goal of an experiment: establishing that the difference in treatments *caused* a difference in responses. This is a key advantage of experiments over observational studies; well designed experiments allow researchers to make cause-and-effect conclusions. An observational study comparing two or more groups—even one involving random selection of individuals from the corresponding populations of interest—cannot provide convincing evidence of causation. Why not? Because we can't isolate the effects of the variable(s) we're interested in from the effects of other variables. We discuss this limitation of observational studies in detail in the final paragraph of the Overview using the well-known setting of trying to determine whether smoking cigarettes causes cancer in humans.

When a well-designed experiment reveals differences in responses between treatment groups, there are two possible explanations: (1) the difference in responses was caused by the different effects of the treatments, or (2) the treatments actually have the same effect on experimental units, so the difference in responses is not due to the effects of the treatments, but rather to the chance involved in the random assignment of treatments to experimental units. More experienced users of statistics can calculate the probability (chance) of obtaining a difference in responses as large as or larger than the one actually observed in the study just from the random assignment. Based on this probability, we can determine whether explanation (2) is a plausible explanation for the observed difference. If not, we conclude that the observed difference is **statistically significant** and that we favor explanation (1). Such decisions based on probability are the foundation of inference, which is introduced in Section IV of this module.

Once students have read about and discussed the three essential experimental design principles—random assignment, control, and replication—in the context of the weed killer example, you may want to ask them to explain how these principles apply in the subsequent example describing the Physicians' Health Study.

The Physicians' Health Study is a famous example of a well-designed experiment that showed taking aspirin regularly helps reduce the risk of heart attack—at least for middle-aged, male physicians.

For reference, here is a complete listing of the vocabulary from the Overview:

> **Experimental units/subjects:** The individuals who take part in an experiment
>
> **Treatments:** The specific conditions that researchers impose on experimental units
>
> **Confounding:** When it is impossible to separate the effects of the treatments from the effects of another variable on the response variable in an experiment
>
> **Random assignment:** A fundamental principle of experimental design that involves using a chance mechanism to allocate treatments to experimental units
>
> **Explanatory variable/factor:** A variable that is deliberately manipulated by the researcher to measure experimental units' responses
>
> **Response variable:** A variable that measures experimental units' responses to the treatments

Control: An important principle of experimental design that entails trying to ensure that variables other than the explanatory variable(s) have roughly equivalent effects on the experimental units that are assigned to the different treatment groups. Researchers can either try to hold the values of such variables constant throughout the experiment or rely on the random assignment to balance out the effects of these variables on the experimental units in different treatment groups.

Replication: A fundamental principle of experimental design that involves giving each treatment to enough experimental units so that any difference in the overall effects of the treatments can be detected

Placebo: A fake treatment

Double-blind: When neither the subjects nor the individuals measuring subjects' responses know who is receiving which treatment

Single-blind: When either the subjects or the people measuring subject's responses, but not both, are unaware of who is receiving which treatment

Statistically significant: A difference in responses that cannot be accounted for by the chance involved in the random assignment of treatments to experimental units

Possible Extensions

You might want to show students a video clip describing the Physicians' Health Study experiment. The Annenberg/Corporation for Public Broadcasting web site, *www.learner.org*, houses a series of instructional statistics videos called "Against All Odds: Inside Statistics." By completing a free registration process, you can play any of these videos as streaming downloads on your computer. The Physicians' Health Study clip is in Video 12: Experiments. The Physicians' Health Study web site, *phs.bwh.harvard.edu*, contains additional information about the experiment, including the results of the beta carotene treatment (no statistically significant difference from placebo beta carotene).

The more recent Women's Health Initiative (WHI), begun in 1991, included clinical trials and an observational study that examined the effects of hormone therapy, diet, and vitamin supplements in postmenopausal women. The WHI's web site is *www.nhlbi.nih.gov/whi*.

Section III: Experiments

Corresponds to pp. 62-66
in Student Module

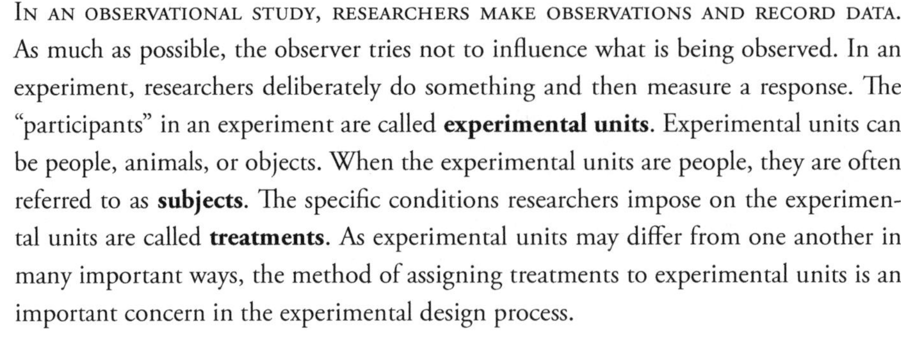

IN AN OBSERVATIONAL STUDY, RESEARCHERS MAKE OBSERVATIONS AND RECORD DATA. As much as possible, the observer tries not to influence what is being observed. In an experiment, researchers deliberately do something and then measure a response. The "participants" in an experiment are called **experimental units**. Experimental units can be people, animals, or objects. When the experimental units are people, they are often referred to as **subjects**. The specific conditions researchers impose on the experimental units are called **treatments**. As experimental units may differ from one another in many important ways, the method of assigning treatments to experimental units is an important concern in the experimental design process.

Let's look at an example. A biologist would like to determine which of two leading brands of weed killer is less likely to harm the broad-leafed plants in a garden at the university. Before spraying near the plants in the garden, the biologist decides to conduct an experiment that will allow her to compare the effects of these two brands of weed killer on broad-leafed pansy plants (one of the varieties in the garden). The biologist obtains 24 individual pansy plants to use in the experiment. In this simple experiment, the *experimental units* are the individual pansy plants and the *treatments* are the two brands of weed killer.

Consider the following two plans for assigning treatments to the pansy plants:

Plan A: Choose the 12 healthiest looking pansy plants. Apply brand X weed killer to all 12 of those plants. Apply brand Y weed killer to the remaining 12 pansy plants.

Plan B: Choose 12 of the 24 individual pansy plants at random. Apply brand X weed killer to those 12 plants and brand Y weed killer to the remaining 12 plants.

Which plan seems preferable? Let's evaluate what might happen with each of these plans.

Under Plan A, suppose the pansy plants treated with brand Y weed killer have many more dead or dying leaves than the pansy plants treated with brand X. Can the biologist feel confident recommending brand X to the campus gardener as the safer weed killer? Not at all. Since the healthier plants received the brand X treatment and the less healthy plants received the brand Y treatment, it could be that more leaves were dead or dying on the pansy plants treated with brand Y because those plants were less healthy to begin with. We really can't separate the effects of the two brands of weed killer from the effect of the original healthiness of the plants in the two groups. The inability to separate the effects of the treatments from the effects of another variable in a study is known as **confounding**.

With Plan B, individual pansy plants are assigned at random to one of the two weed killer treatments. This **random assignment** helps to ensure that the group of plants treated with brand X and the group of plants treated with brand Y are fairly similar to begin with in terms of all characteristics that might affect the plants' responses to the treatments. If the biologist then observes that the pansy plants treated with brand Y weed killer have many more dead or dying leaves than the pansy plants treated with brand X, there are two plausible explanations for the observed difference.

First, it is possible that there is no difference in the effects of the two brands of weed killer on pansy plants. Some pansies are heartier than others, and, just by chance, the random assignment placed more of those healthy plants in the group that was treated with brand X. In other words, the observed difference could be simply due to chance.

The second possible explanation is that brand X weed killer actually results in greater harm to pansy plants than brand Y. In that case, we could say the difference in the number of dead or dying leaves between the two groups of pansy plants is a direct result of the brand of weed killer used. Put another way, the difference in brand of weed killer *caused* the difference in the number of dead or dying leaves.

Random assignment of treatments to subjects is an essential component of well-designed experiments. One of the big advantages of such experiments is their ability to help the researcher establish that changes in one variable (like brand of weed killer) cause changes in another variable (like number of dead or dying leaves). Since establishing causation is often a goal of experiments, we find it useful to give names to the two variables mentioned in the previous sentence. We call the variables that the experimenters directly manipulate the **explanatory variables** or **factors** and the variables that measure the subjects' responses to the treatments the **response variables**. The treatments in an experiment correspond to the different possible values of the explanatory variables. For the weed killer experiment above, there is one factor—brand of weed killer—and one response variable—number of dead or dying leaves.

In addition to randomly assigning treatments to experimental units, there are two other important considerations in designing experiments. The first is to **control** for the effects of variables that are not factors in the experiment but that might affect experimental units' responses to the treatments. Some variables can be controlled by trying to keep them at a constant value. For example, the biologist would want to ensure that the plants all receive the same amount of water and are exposed to the same amount of light. If everything is roughly equivalent for the two groups of plants except for the treatments, and we observe a difference in the response variable, then that difference is either a result of the random assignment or is caused by the difference in treatments.

Some variables can't be easily controlled by keeping them at a constant value. One such variable in the weed killer example was the current state of health of the plant. In this case, the random assignment of plants to treatments should help spread the healthy and less healthy plants out in a fairly balanced way between the two groups of pansy plants. Then, any differences in the number of dead or dying leaves that appear should not be a result of differences in initial plant health.

The other important experimental design principle is **replication**. In a nutshell, replication means giving each treatment to enough experimental units so that any difference in the effects of the treatments is likely to be detected. Imagine the biologist treating one pansy plant with brand X weed killer and one pansy plant with brand Y weed killer. If the plant treated with brand Y has more dead or dying leaves, can the biologist conclude that brand X is safer to use on the university's pansy plants? Of course not. Individual pansy plants vary widely in terms of general health and other characteristics that might affect their response to a particular brand of weed killer. With only one experimental unit available for each treatment, the random assignment can't be counted on to produce roughly "equivalent" groups prior to administering the treatments. Any difference we observe in the number of dead or dying leaves on the two pansy plants could simply be due to the difference in the initial health of the plants.

Now imagine the biologist conducting the same weed killer experiment, but with 50 pansy plants receiving each treatment. If the pansies treated with brand Y have a much higher number of dead or dying leaves than the pansies treated with brand X, the biologist should feel much more confident concluding that the difference in treatments caused the observed difference in the response variable.

Let's look at one more example. In the fall of 1982, researchers launched a now famous experiment investigating the effects of aspirin and beta carotene on heart disease and cancer. Over 22,000 healthy male physicians between the ages of 40 and 84 agreed to serve as *subjects* in the experiment. The two *factors* being manipulated by the researchers were whether a person took aspirin regularly and whether a person took beta carotene regularly. Researchers decided to use four treatments: (1) aspirin every other day and beta carotene every other day, (2) aspirin every other day and "fake" beta carotene every other day, (3) "fake" aspirin every other day and beta carotene every other day, and (4) "fake" aspirin every other day and "fake" beta carotene every other day.

The "fake" pills looked, tasted, and smelled like the pills with the active ingredient, but had no active ingredient themselves. (We call such "fake" treatments **placebos**.) Subjects were randomly assigned in roughly equal numbers to the four groups. Several *response variables* were measured in the study, including whether the individual had a heart attack and whether the individual developed cancer. Neither the subjects nor the people measuring the response variable knew who was receiving which treatment. We say this experiment was carried out in a **double-blind** manner. If either the subjects or the people measuring the response variable knows who is receiving which treatment, but the other doesn't, then the experiment is **single-blind**.

An outside group of statisticians that was monitoring the Physicians' Health Study reviewed data from the experiment on a regular basis. To everyone's surprise, the data monitoring board stopped the aspirin part of the experiment several years ahead of schedule. Why? Because there was compelling evidence that the subjects taking aspirin were having far fewer heart attacks than those who were taking placebo aspirin. It would have been unethical to continue allowing some physicians to take a placebo with clear evidence that aspirin reduced the risk of heart attack.

Even though the Physicians' Health Study was an exceptionally well-designed experiment, it does have some limitations. Researchers decided to use male physicians as subjects because they felt doctors would be more likely to understand the importance of taking the pills every other day for the duration of the study. That may be true, but because only male physicians were used in the study, we cannot generalize the findings of this study to women, or even to all male adults. We can feel pretty confident concluding that taking aspirin regularly *caused* a reduction in heart attack risk. However, the benefits of taking aspirin regularly might be offset by other effects of the drug, such as an increased risk of stroke. In spite of its limitations, the Physicians' Health Study provided a template for other researchers who wanted to design experiments to help answer important questions.

In many published reports of experimental studies, we see conclusions such as "the observed difference in heart attack rates was **statistically significant**." This tells us that the differences in the response variable between those in different treatment groups cannot reasonably be explained by the chance involved in the random assignment of treatments to subjects. Recall what we said earlier: There are only two possible explanations for the observed differences in an experiment—that they were due to the chance involved in the random assignment or that the difference in treatments caused the difference in the response variable. Saying that the results of a particular experiment are *not* statistically significant means that we can't rule out the possibility that there is no difference in the effects of the treatments, and that the differences in response are simply due to the random assignment.

You may have noticed that in both the examples presented here, the subjects were *not* randomly selected from a larger population. This is usually the case with experiments. It often isn't practical to choose subjects at random from the population of interest. Consider how you would go about randomly selecting 24 pansy plants from the population of *all* pansy plants, for example. Or how researchers might randomly select 22,000 male physicians. As you learned earlier, the lack of random selection limits our ability to generalize results to the population of interest.

However, even if experimental units are not randomly selected, well-designed experiments can give convincing evidence that changes in one variable cause changes in another variable. Establishing causation is much more difficult with observational studies, because researchers cannot hold other variables constant and cannot assign individuals at random to treatment groups. As an example, consider early observational studies that suggested people who smoked were much more likely to get lung cancer than people who didn't smoke. Cigarette company executives argued that *confounding* was at work. They claimed that the kinds of people who smoked were also much more likely to engage in other unhealthy activities—such as drinking, overeating, and failing to exercise—than people who didn't smoke. It was these other unhealthy behaviors, they said, that led to increased risk of cancer, not smoking cigarettes. After many other observational studies showed the strong connection between smoking and lung cancer, and experiments on animal subjects demonstrated that smoking caused cancerous growths, cigarette company executives finally conceded.

Teacher Notes for Investigation #10: Do Diets Work?

In this investigation, students will review and critique two experiments designed to compare the effectiveness of low-carbohydrate and low-fat diets in reducing weight and cholesterol in obese adults.

Prerequisites

Students should be able to:

Identify the subjects, factor(s)/explanatory variable(s), treatments, and response variable(s) in an experimental setting

Distinguish an observational study from a survey or an experiment

Explain the purpose of randomly assigning treatments to subjects in an experiment

Determine whether an experiment was carried out in a single-blind or double-blind manner

Explain the purpose of control in an experimental design

Explain what is meant by "statistical significance"

Identify a potential confounding variable in a study and explain how the variable could result in confounding

Explain how the way in which data were produced affects our ability to generalize results to a larger population of interest

Learning Objectives

As a result of completing this investigation, students should be able to:

Explain how the design principle of control applies in a specific experimental setting

Interpret experimental results in context

Explain what it means for a result to not be statistically significant in the context of an experiment

Describe possible limitations of an experiment, such as side effects and dropouts

Summarize and critique an experiment based on written information about the experiment

Teaching Tips

One of the primary goals of this first investigation in the Experiments section is to increase students' familiarity with and comfort in applying the terminology of experiments. Students may want to refer to the Overview as they complete the investigation.

Be sure to discuss how data ethics apply in these experimental settings: informed consent, anonymity and confidentiality, and external review board.

We recommend having students work through the questions in pairs initially. The questions are divided into four distinct groups. Questions 1 through 5 focus on the design of the two experiments. Questions 6 through 8 ask students to draw preliminary conclusions about low-carb versus low-fat diets based on the results of these two studies.

Questions 9 through 12 address some possible limitations of these experiments. Finally, Questions 13 and 14 ask students to refine their preliminary conclusions in light of the possible limitations.

To promote effective communication, you may want to have students discuss their responses with a partner prior to sharing answers with the class. You might also ask students to provide feedback on each other's answers in a whole class setting before you evaluate the accuracy and clarity of their responses.

Suggested Answers to Questions

1. The completed table is shown below.

	Duke University Study	**Philadelphia Study**
Subjects	120 volunteers, aged 18 to 65, with high cholesterol	132 obese adult volunteers
Factor(s)/ explanatory variable(s)	Type of diet followed	Type of diet followed
Treatments	Low-carb, high-protein diet	Low-carbohydrate diet
	Low-fat, low cholesterol diet	Low-fat diet
Response variable(s)	Change in weight	Change in weight
	Change in cholesterol	Change in cholesterol

2. In both the Duke University study and the Philadelphia study, researchers deliberately imposed treatments—either a low-carbohydrate diet or a low-fat diet—on the subjects. When something is deliberately done to individuals in a study to measure their responses, the study is an experiment.

3. Researchers assigned subjects at random to either a low-fat or low-carbohydrate diet. By letting chance divide the available subjects into two groups, the researchers were attempting to ensure the groups were roughly equivalent in terms of variables other than the specific diets assigned that might affect subjects' responses to the treatments. The researchers were also trying to avoid any bias that might have resulted from subjectively assigning subjects to treatment groups.

4. These experiments could have been conducted in a single-blind manner if the individuals who interacted with the subjects and measured the response variables did not know who was assigned to each of the diet treatments. As the subjects would know what kinds of foods they were eating, it would not have been possible to carry out either experiment in a double-blind fashion.

5. (a) If any of the subjects had dieted recently, their bodies might have responded differently to the diet regimens assigned in the Duke experiment than if they had not been dieting. Likewise, subjects who had used weight loss medications during the previous six months might have responded differently to the diet treatments assigned in the Duke study as a result of lingering effects of those medications. By using only subjects who had not dieted or used weight loss medications in the previous six months, researchers attempted to control for the effects of other variables that might have systematically affected subjects' responses to the diet treatments.

(b) Because exercise could affect subjects' weight loss and change in cholesterol level, it was important for researchers to try to ensure that all participants in the experiment engaged in similar amounts of exercise. Otherwise, any differences in weight loss or cholesterol level between the two groups of subjects could have been the result of differing exercise habits, rather than the specific diets assigned to those groups.

6. Both experiments suggest that following a low-carbohydrate diet caused a greater decrease in weight over a six-month period than following a low-fat diet. Likewise, both experiments suggest that following a low-carb diet caused a greater increase in HDL (good) cholesterol than following a low-fat diet. The Philadelphia experiment did not show a significant difference in weight loss for subjects on a low-carb diet when compared to those on a low-fat diet over a one-year period. So it is possible that a low-carb diet is more effective at reducing weight in the short-run than a low-fat diet, but that the two diet regimens result in similar amounts of weight loss over longer periods of time. One important caveat: These conclusions only apply to individuals like those who were willing to take part in these two experiments—somewhat motivated, otherwise healthy, obese adults.

7. This difference in average weight loss (2 kg) for subjects in the two groups was not large enough to rule out the possibility that the observed difference was simply due to the luck of the random assignment, and not to the effects of the two diet treatments.

8. Although the low-carb diet showed significant benefits in terms of weight loss and decrease in cholesterol over a six-month period, it also resulted in more minor side effects, such as constipation and headaches, than did the low-fat diet.

9. With such a high dropout rate in both experiments, our conclusions would be open to challenge. Researchers don't know what would have happened to the subjects who dropped out in terms of weight loss or change in cholesterol level. It is possible that the results of the experiment would have been different if all the subjects had participated for the full duration of the study. We have no way of knowing in what way the results might have differed.

What if most of the dropouts in the Philadelphia study had been from the low-carb diet group? Maybe those people withdrew from the study because they weren't experiencing a decrease in weight loss. If that was the case, then had those subjects remained in the experiment for the entire six months, researchers might not have observed a significant difference in weight loss for the two diet treatments. The fact that a much higher percentage of subjects in the low-fat diet group than of subjects in the low-carb diet group dropped out of the Duke University experiment is concerning.

Researchers should follow up with individuals who drop out of an experiment to find out why they made that decision.

10. If subjects did not follow their assigned diet treatments, then the results of the experiment are no longer as convincing. Researchers are drawing conclusions based on the belief that subjects are following their assigned diet plans. If some subjects deviate from the assigned diet regimen, researchers can no longer attribute any significant differences in weight loss or cholesterol level to the difference in the diet treatments.

11. As the daily nutritional supplement represents another systematic difference between the two groups of subjects (in addition to the diet plan they're following), researchers would need to rule out the possibility that differences in the response variables between the two groups could be due to the daily nutritional supplement and not the low-carb or low-fat diet.

12. In the Duke University study, a potential confounding variable is whether subjects took a daily nutritional supplement. To be potentially confounding, the variable must be associated with group membership and have an effect on the response variables. Since only the subjects in the low-carb diet group took the daily nutritional supplement, there is a clear association between this variable and group membership in the experiment.

As another example, consider the variable "amount of exercise." Amount of exercise could clearly affect weight loss or change in cholesterol level. In order for this to be a potential confounding variable, however, it would also have to be the case that subjects in one group tended to exercise more than subjects in the other group. As researchers randomly assigned subjects to the two diet treatments, the groups should have started out fairly balanced in terms of exercise habits.

13. No. The subjects who participated in both these experiments were recruited to do so. That is, they were willing volunteers. Perhaps these individuals were more motivated to begin with than the general population of obese adults. Also note that the subjects in both experiments were obese adults. Consequently, the results of the experiments apply only for otherwise healthy, obese adults, not to overweight adults in general. We can only generalize the findings of these two experiments to a population of individuals like the subjects who actually participated.

14. Answers will vary. Students should include the following points in their summaries:

Both experiments suggested a low-carb diet resulted in greater weight loss over a six-month period than did a low-fat diet.

The Philadelphia experiment found no significant difference in weight loss between the low-fat diet and the low-carb diet over a one-year period. The 2 kg difference in average weight loss that researchers observed could have been due to the random assignment of subjects to groups, and not due to the difference in diet regimens.

Both experiments suggested that a low-carb diet resulted in a significantly higher increase in LDL (good) cholesterol than a low-fat diet.

The high dropout rates in both experiments are concerning. We don't know how the results would have been affected if these subjects had completed the experiment.

In the Duke experiment, subjects in the low-carb group were given a daily nutritional supplement, but those in the low-fat group weren't. This is a potential source of confounding.

Researchers can only generalize the results of these experiments to the population of otherwise healthy, obese adults like the ones who agreed to participate in these studies.

Possible Extensions

You might want to have students find an article describing the results of another experiment on dieting and weight loss, and then have them perform an analysis similar to the one outlined in this investigation.

Investigation #10: Do Diets Work?

The Atkins Diet is one of many popular weight loss diets. It is based on reducing the consumption of carbohydrates. For years, such "low-carb" diets have been touted as being effective for weight loss and other health benefits. But before 2001, no one had attempted to demonstrate the effectiveness of a low-carb diet in a well-designed comparative experiment. Then, two separate groups of researchers attempted to do just that.

Corresponds to pp. 67-71 in Student Module

At Duke University Medical Center, Dr. William Yancy and his colleagues recruited 120 people between the ages of 18 and 65. All of the participants were obese and had high cholesterol, but were otherwise in generally good health. Researchers randomly assigned half of the participants to a low-carbohydrate, high-protein diet (similar to an Atkins Diet) and the other half to a low-fat, low-cholesterol diet. At the end of six months, researchers measured the change in each participant's weight and cholesterol levels.[1]

In the second study, Dr. Linda Stern and her colleagues recruited 132 obese adults at the Philadelphia Veterans Affairs Medical Center in Pennsylvania. Half of the participants were randomly assigned to a low-carbohydrate diet and the other half were assigned to a low-fat diet. Researchers measured each participant's change in weight and cholesterol level after six months and again after one year.[2]

1. Complete the following table using the details provided above about the two studies.

	Duke University Study	**Philadelphia Study**
Subjects		
Factor(s)/Explanatory Variable(s)		
Treatments		
Response Variable(s)		

---

1 "A Low-Carbohydrate, Ketogenic Diet versus a Low-Fat Diet To Treat Obesity and Hyperlipidemia," by Yancy, William S. et al, *Annals of Internal Medicine*, May 2004, 140(10) 769-777.

2 "The Effects of Low-Carbohydrate versus Conventional Weight Loss Diets in Severely Obese Adults: One-Year Follow-up of a Randomized Trial," by Stern, Linda et al, *Annals of Internal Medicine*, May 2004, 140(10) 778-785.

2. Explain why both of these studies are experiments, and not observational studies or surveys.

3. How did the researchers in both studies determine which subjects received which treatments? Why did they use the method they did?

4. Could these experiments have been carried out in a single-blind or double-blind manner? Justify your answer.

5. Each of the following quotations describes the subjects in the Duke University experiment. Explain how each is an example of control and why it is important in terms of the design of the study.

(a) "None had dieted or used weight loss medications in the previous six months."

(b) "All subjects were encouraged to exercise 30 minutes at least three times per week and had regular group meetings at an outpatient research clinic for six months."

Let's look at some results from the two studies.

> In the Duke University experiment, over the six-month duration of the study, weight loss was 12.9% of original body weight in the low-carbohydrate diet group and 6.7% of original body weight in the low-fat diet group. The low-carb diet group showed a greater increase in HDL (good) cholesterol than the low-fat diet group.

> In the Philadelphia experiment, subjects in the low-carbohydrate diet group lost significantly more weight than subjects in the low-fat diet group during the first six months of the study. At the end of a year, however, the average weight loss for subjects in the two groups was not significantly different. The low-carbohydrate diet group did show greater increase in HDL (good) cholesterol level after a year than the low-fat diet group.

6. Briefly summarize what the results of these two experiments seem to suggest about the relative effectiveness of low-carbohydrate diets and low-fat diets on weight and cholesterol.

7. In the Philadelphia experiment, the subjects in the low-carbohydrate diet group lost an average of 5.1 kg in a year. The subjects in the low-fat diet group lost an average of 3.1 kg. Explain how this information could be consistent with the statement above about the average weight loss in the two groups not being significantly different.

8. Here is an excerpt from a report about the Duke University experiment: "Participants in the low-carbohydrate diet group had more minor adverse effects, such as constipation and headaches, than did patients in the low-fat diet group." How would you modify your summary in question 6 based on this additional information?

When you look at experimental results, it's important to consider possible limitations of the study. The next few questions will help you look critically at the two experiments described earlier.

9. Explain how the following excerpts from a report about the two experiments might affect your conclusions about the effectiveness of low-carb versus low-fat diets:

Duke University study: "The study was completed by 76% of participants in the low-carbohydrate diet group and by 57% of participants in the low-fat diet group."

Philadelphia study: "Study limitations include high dropout rate of 34% …"

10. In both experiments, participants were assigned at random to a low-fat or low-carbohydrate diet group. What exactly does that mean? The subjects in the low-fat diet group attended counseling sessions about how to restrict their caloric intake from fat. The subjects in the low-carbohydrate group attended counseling sessions about how to restrict their carbohydrate intake. These counseling sessions continued on a weekly or monthly basis throughout the experiment. It is possible that some people in each group did not restrict their diets as instructed. How might this affect conclusions based on the experiment?

11. In the Duke University study, subjects in the low-carbohydrate group all received daily nutritional supplements. Subjects in the low-fat group did not. How might this affect conclusions based on the experiment?

12. Give an example of a potential confounding variable in one of the two experiments. Explain carefully how the factor you choose could result in confounding.

13. Is it reasonable to generalize the results of these two experiments to the population of all overweight adults? Justify your answer.

14. Now that you have considered possible limitations of these two experiments, summarize what the results of these two experiments seem to suggest about the relative effectiveness of low-carbohydrate diets and low-fat diets on weight and cholesterol. You may want to refer to what you wrote earlier in response to question 6.

Teacher Notes for Investigation #11: Distracted Learning

IN THIS INVESTIGATION, STUDENTS WILL DESIGN, CARRY OUT, AND ANALYZE RESULTS from an experiment to determine whether listening to Mozart while performing a memorization task helps students remember better than doing a similar task with no music playing.

Prerequisites

Students should be able to:

Explain how the way in which data were produced affects our ability to generalize results to a larger population of interest

Identify a potential confounding variable in a study and explain how the variable could result in confounding

Explain the purpose of randomly assigning treatments to subjects in an experiment

Carry out the random assignment of treatments to subjects in an experiment

Identify the subjects, factor(s)/explanatory variable(s), treatments, and response variable(s) in an experimental setting

Construct and interpret a comparative dotplot for a quantitative variable, describing shape, center, spread, and any unusual values

Construct and interpret a dotplot of differences for paired data, describing shape, center, spread, and any unusual values

Choose the most appropriate numerical measures of center and spread to use in a given setting (mean and standard deviation OR median and interquartile range [IQR])

Determine whether an experiment was carried out in a single-blind or double-blind manner

Learning Objectives

As a result of completing this investigation, students should be able to:

Consider alternative designs for an experiment, and then choose the best one for answering a given research question

Explain why it is important for the order of treatments to be randomly assigned to subjects in a design that requires each subject to receive both treatments

Draw appropriate conclusions from an experiment involving paired data from volunteer subjects

Make at least one suggestion for improving the design of an experiment based on the actual experience of carrying out that experiment

Teaching Tips

Questions 1 through 4 of this investigation walk students through the process of designing an experiment to test whether listening to Mozart improves memorization skills for students in their class. Students are steered away from the design used in the two experiments of the previous investigation, in which subjects were randomly assigned into two roughly equal treatment groups. This type of design is known as a *completely randomized design*. Instead, students are nudged toward using a *matched pairs design*, in which each subject receives both treatments in a random order.

Why is a matched pairs design preferable in this case? We know that individuals vary widely in their memorization abilities. If we used a completely randomized design, with about half the students in the class assigned to the Mozart treatment and the other half assigned to work in silence, we would expect considerable variation in the individual scores on the memorization task within each group. If we observe a difference in the mean scores for the two groups, we would like to know whether that difference was caused by listening to Mozart. Of course, there is another possible explanation for any difference that emerges. Maybe subjects would perform the same whether they listened to Mozart or not, so the observed difference is simply a result of which subjects were randomly assigned to each group. With lots of variation present, it will be more difficult to rule out this second possible explanation in favor of a causal connection between listening to Mozart and memorization.

By using a matched pairs design, we isolate the variation among individuals by comparing each individual's performance on two similar memorization tasks—one while listening to Mozart and one done in silence. We perform our analysis on the difference in memorization scores for the students in the class. There should be less variation in the difference values than there would have been with data produced using a completely randomized design. As a result, it should be easier to detect a "Mozart effect" if there is one by ruling out chance variation from the random assignment as a plausible explanation.

Question 5 asks students to review the details of their design before implementing it. In Question 6, students carry out the random assignment for their design. In Question 7, students actually perform the experiment. Here are two memorization tasks that students can use:

Task A: 12 09 96 62 66 52 26 82 25 18 98 31 06 48 47 72 28 67 85 57

Task B: 38 07 18 85 73 90 31 12 37 39 87 33 06 44 43 34 08 27 24 99

Questions 8 through 16 take students through the process of analyzing data, identifying possible limitations, and drawing conclusions. If students use a matched pairs design for their experiment, it would be inappropriate for them to analyze the "with Mozart" and "in silence" data as if they came from two unrelated groups of individuals, as Question 8 seems to suggest. Make the point that **the appropriate method of data analysis is determined by the design of the study**. If data are paired by design, then students should analyze the pairs of data values. In this case, that means examining the differences in performance scores for the subjects.

Suggested Answers to Questions

1. Since the subjects are available volunteers, and are not randomly selected from a larger population of interest, we will only be able to generalize our findings to the population of students that are similar to the people in this class.

2. With this design, the two groups of subjects would be performing the experiment in two different locations. It is possible that students will perform differently on the task as a result of the conditions in the two rooms. If so, then "room conditions" would be a confounding variable. The process of relocating to another room may affect the subject's performance on the task in a systematic way. Perhaps the movement will stimulate these students' brains, resulting in better performance on the memorization task than for those students who stay put. Because individuals vary widely in their ability to memorize, it might be better to have each subject perform a similar memorization task twice—once while listening to Mozart and once in silence—so that individual differences in memorization are planned for, rather than distributed between, the two groups with random assignment. After all, the random assignment could lead to two groups with large amounts of variability in their memorization skills, which would make it more difficult to detect any effect of listening to Mozart on memorization.

3. (a) By separating the "good" and "not-so-good" memorizers in advance based on performance on the initial memory task, we would expect less variability in memorization abilities for the randomly assigned groups of subjects in each performance category than for the two randomly assigned groups in the design proposed in question 2. With less variability present, it should be easier to detect any effect of listening to Mozart on memorization for "good" memorizers and for "not-so-good" memorizers.

(b) Have each subject perform two similar memorization tasks, one while listening to Mozart and one in silence. Randomly assign the subjects into two approximately equal groups. Have one group do the first task while listening to Mozart and the second task in silence. Have the other group do the first task in silence and the second task while listening to Mozart.

4. (a) Even if the two memorization tasks are similar, subjects may still find one task more difficult than the other. Suppose the subjects find Task A easier than Task B. If subjects perform better while listening to Mozart, it might be because they are doing the easier task, and not because of the music. In other words, which task subjects perform would be a potential confounding variable.

(b) Students may learn from doing the first memorization task, and perform better on the second memorization task as a result. This is known as a *learning effect*. In this scenario, if students performed better while listening to Mozart, we wouldn't know whether this was due to a learning effect or due to the effects of the music.

(c) Students should design a method of random assignment in which about equal numbers of students will perform the experiment under each of the following four conditions:

Task A with Mozart, then Task B in silence

Task A in silence, then Task B with Mozart

Task B with Mozart, then Task A in silence

Task B in silence, then Task A with Mozart

(d) Answers will vary, depending on the random assignment plan that was agreed upon in (c). One method would be to have students write their names on roughly identical slips of paper, put the slips in a hat, and mix them thoroughly. Then, you could draw out names one at a time, with the first person assigned to the first set of experimental conditions from (c), the second person to the second set of experimental conditions from (c), and so on. Of course, students could use a variation of the hat method by assigning distinct numbers to the members of the class, and then using a random digit table or random number generator to mimic the process described in the previous sentence.

Students could opt to roll a four-sided die (or a six-sided die, ignoring two of the numbers) for each member of the class to determine which of the four experimental conditions from (c) that person would follow. Note that this method could result in somewhat unequal numbers of students following each of the four experimental conditions just by chance.

5. (a) The students in this class.

(b) The explanatory variable is what a person listens to while performing a memorization task.

(c) Treatments are connected with values of the explanatory variable. In this case, the two possible values of the explanatory variable are "listen to Mozart" and "work in silence." The two treatment combinations for our experiment are (1) listen to Mozart during the first task; work in silence during the second task, and (2) work in silence during the first task, then listen to Mozart during the second task. We're going to measures students' performances on the tasks as part of the experiment. However, the tasks themselves are not treatments, because we are not deliberately imposing the tasks on the students to measure their responses to those tasks.

(d) A scoring system such as one point for each number remembered correctly, and minus one point for each incorrect number that is listed, might be a good way to measure performance and avoid haphazard guessing.

(e) The response variable is the difference in score on the memorization tasks while listening to Mozart and while working in silence.

6. Answers will vary.

7. Data will vary!

8. Comparative dotplots will vary. Note that the horizontal axis in the plot represents the score on the memorization task, which could be positive, negative, or zero based on the scoring system that was suggested in 5(d). When describing similarities and differences, students should discuss issues of shape, center, spread/variability, and any unusual values.

9. Difference values will vary.

10. Dotplots will vary. Note that the horizontal axis in the plot represents the difference in score on the two memorization tasks for each student. Since students are testing the belief that Mozart might help improve memorization, they might want to define difference = score with Mozart − score in silence. When interpreting the plot, students should discuss issues of shape, center, spread/variability, and any unusual values in the context of this experiment.

11. The dotplot in question 10 shows the difference in score for each student when listening to Mozart versus when performing the memory task in silence. The dotplot in question 8 treated the two scores for each student as unrelated values, simply showing all students' memorization scores with Mozart and all students' memorization scores without Mozart. Because the two scores for each student are related (by virtue of being produced by the same individual), it is more appropriate to focus on the difference in scores when making a graphical display of the data. The plot in question 10 makes it easier to see whether listening to Mozart helped increase memorization performance for students in the class, which was the goal of the experiment.

12. Answers will vary. Students could use the mean and standard deviation to summarize center and spread, respectively, if the distribution of differences is roughly symmetric and there are no potential outliers. If the distribution is clearly skewed, or potential outliers are present, then the median and interquartile range (IQR) would be more appropriate summaries of center and spread.

13. This experiment was neither single-blind nor double-blind. Both the subjects and the individuals measuring the response variable (memorization score) knew which treatment combination the students were receiving.

14. Answers will vary. Students should be evaluated on how well they use the evidence from their graphs and numerical summaries to support their answer.

15. No. We can only generalize our findings about listening to Mozart to memorization tasks that are similar to the ones used in this experiment.

16. Having each student listen to the same Mozart selection was a form of control. It is possible that students would respond differently to other Mozart pieces or other kinds of music when performing similar memorization tasks. Consequently, we can't generalize the results of this study to all Mozart tunes or other types of music.

Possible Extensions

There are plenty of possible variations on this experiment that students could design and carry out. For instance, the original claim of researchers who discovered the so called "Mozart effect" was that listening to Mozart helps improve performance on spatial reasoning tasks. Students could use mazes as the task, rather than lists of numbers to memorize.

 Investigation #11: Distracted Learning

While you study, do you watch TV, listen to music, check your MySpace page, surf the Internet, chat on e-mail, talk or text on your cell phone? Do your parents insist that you can't possibly concentrate on studying while you're distracted by one of these activities? Maybe the conversation goes something like this:

Parent: "Take off your headphones and do your homework!"

Student: "I am doing my homework, and I work better with my music on."

Parent: "Turn it off! You can't study with that distraction!"

Student: "Yes I can. It helps me relax."

Parent: "Turn off that racket and concentrate on your school work!"

Student: "I study better with it on!"

Corresponds to pp. 72-79 in Student Module

Who is right? Some say that any distraction might interfere with your focus on the work you're doing, which may in turn affect the quality of the finished product. But others argue that listening to music actually helps them concentrate because the music "drowns out" other potential distractions. What do you think? Can previous research help us sort this out?[1]

In 1993, Frances Raucher and his colleagues designed an experiment to test whether listening to Mozart would help students improve their performance on a spatial reasoning task. They recruited 36 college students to participate in the experiment. The subjects were randomly assigned to three groups, with 12 students per group. Subjects in Group 1 listened to a 10-minute selection from a Mozart piece. Group 2 listened to a relaxation tape for 10 minutes. Subjects in Group 3 sat in silence for 10 minutes. Each subject took a pretest on spatial reasoning two days before the experiment and a post-test on spatial reasoning immediately after the 10-minute treatment. The results of the experiment seemed surprising: Students who listened to Mozart showed significantly higher gains in their scores on spatial-reasoning tasks than students in the other two groups.

After hearing the results of Rauscher's experiment, some eager parents started playing Mozart tapes for their children in hopes of increasing their spatial reasoning skills. One state even passed legislation requiring preschools to play 30 minutes of classical music a day. Other researchers tried to confirm this so-called "Mozart effect" in experiments of their own, but with little success.

So the question remains: Does listening to music help or hinder students' learning? The answer may depend on what type of "learning" we mean. In this investigation, your class will design and carry out an experiment to test whether listening to music helps or

---

1   *www.madsci.org/posts/archives/mar98/889467626.Ns.r.html* served as inspiration for part of this investigation.

hinders students as they perform a memorization task. Then, you will analyze data from the experiment and draw some preliminary conclusions from your research.

1. For simplicity, the members of your class will serve as the subjects in your experiment. How might this affect your ability to generalize the results of your study?

2. One possible design for the experiment would be to randomly assign about half of the students in your class to perform the memorization task while listening to Mozart, and the other half to perform the task in a silent room nearby. Then, you could compare the scores of students who listened to Mozart while memorizing with the scores of students who didn't. What flaw(s) do you see in using this design to conduct the experiment?

3. Some people are better at memorizing things than others. Here's another possible design for your experiment that takes this fact into account. Begin by having each student perform a memory task. Based on students' performance on this task, split the class into two roughly equal-sized groups containing the "good memorizers" and the "not-so-good memorizers." Randomly assign about half of the good memorizers to perform a second memory task while listening to Mozart, and the other half to perform the task in a silent room nearby. Use the same random assignment strategy for the not-so-good memorizers. To analyze the data from the experiment, you would compare the change in scores from the first memory task to the second for the good memorizers who listened to Mozart and those who didn't, and separately for the not-so-good memorizers who did and didn't listen to Mozart while memorizing.

(a) In what ways does this design improve on the design from question 2?

(b) How might you further improve the design of this experiment using the idea that some people are better memorizers than others? Explain.

4. Perhaps the best way to take individual differences in memorization skills into account in this experiment is to have each person perform two memory tasks—one while listening to Mozart and one in silence. Then, you can analyze data on the difference in performance for all students in your class and determine whether listening to Mozart seems to help or hurt memorization.

To carry out the experiment in this way, you will need two different but similar memory tasks. Let's call them task A and task B.

(a) Explain why you should not have all students perform task A while listening to Mozart and task B while in a silent room.

(b) Explain why you should not have all students perform their first memory task while sitting in a silent room and their second memory task while listening to Mozart, or vice versa.

(c) Discuss with your classmates how you could use random assignment to most effectively address the issues raised in parts (a) and (b). Once you have settled on a plan, propose it to your teacher.

(d) Describe carefully how you will perform the random assignment required by your approved plan from part (c).

5. Now that we have settled on a design for the experiment, let's confirm some of the details.

(a) Who are the subjects in this experiment?

(b) What factor(s)/explanatory variable(s) is this experiment investigating?

(c) What treatments are being administered? Explain why task A and task B are not treatments.

(d) Let's take a look at the tasks. Each subject will be presented with a list of 20 randomly generated two-digit numbers, such as the list shown below. The student will then have one minute to memorize as many of the numbers in the list as possible. At the end of the minute, each student will have two minutes to write down as many of the numbers as he or she can remember.

| 26 | 86 | 64 | 65 | 75 | 11 | 49 | 47 | 85 | 19 |
| 23 | 57 | 97 | 00 | 62 | 43 | 66 | 94 | 79 | 50 |

A wily student might just write down a bunch of two-digit numbers during the two minute period, hoping to match as many as possible. How might you score performance on this task to reward students for actual memorization and not for guessing?

(e) Based on your answer to (d), describe the response variable(s) this experiment will measure.

Now it's time to do the experiment! Your teacher will assist with logistics so that all students can participate.

6. Carry out the random assignment required for your experiment from question 4(d). Indicate clearly what each student will be doing first and second. You may find it helpful to make a chart like the one below that summarizes how the experiment will be carried out.

Subject	First Task	First Treatment	Second Task	Second Treatment
1	A	Music	B	Silence
2	A	Silence	B	Music
3	B	Music	A	Silence
4	B	Silence	A	Music

Subject	Which Task First? (A or B)	Music First? (Yes or No)	Score With Music	Score Without Music	Difference

7. Have students perform the two memorization tasks as specified in question 6. Record data from the experiment in the table on the previous page.

8. Construct comparative dotplots or boxplots of the scores with music and the scores without music. Describe any similarities and differences you see in a few sentences.

9. Calculate the difference in scores for each student when listening to Mozart versus sitting in a silent room. As a class, decide on which order you will subtract the values. Record these values in the right-most column of the table on the previous page.

10. Construct an appropriate graph of the difference in memorization scores. Describe what the graph tells you in a couple of sentences.

11. In what way is the graph you constructed for question 10 more informative than the comparative graph from question 8?

12. Calculate a measure of center (mean or median) and a measure of spread that you think summarize the differences well. Explain why you chose the measures you did.

13. Was this experiment single-blind, double-blind, or neither? Justify your answer.

14. Based on the results of your experiment, does it appear that listening to Mozart helps or hinders students' performance on memorization tasks? Give appropriate graphical and numerical evidence to support your answer.

15. Can we generalize the results of this experiment to any kind of task that requires memorization? Justify your answer.

16. Why did we have all students listen to the same piece of Mozart music, rather than letting each student choose music he or she liked? Explain.

Teacher Notes for Investigation #12: Would You Drink Blue Soda?

IN THIS CULMINATING INVESTIGATION FOR THE EXPERIMENTS SECTION OF THE MODULE, students will design, carry out, and analyze data from an experiment to test whether people have a preference for blue-colored soda. By this point, students should feel fairly comfortable with the terminology and basic concepts of experimental design. If students completed the previous investigation using a matched pairs design, then they should need little prodding to come up with a similar design for this taste test experiment.

Prerequisites

Students should be able to:

Define a research question

Explain why it is important for the order of treatments to be randomly assigned to subjects in a design that requires each subject to receive both treatments

Carry out the random assignment of treatments to subjects in an experiment

Identify the subjects, factor(s)/explanatory variable(s), treatments, and response variable(s) in an experimental setting

Determine whether an experiment can be carried out in a single-blind or double-blind manner

Explain how the way in which data were produced affects our ability to generalize results to a larger population of interest

Consider alternative designs for an experiment, and then choose the best one for answering a given research question

Use appropriate graphical and numerical techniques for describing the distribution of a categorical variable and for describing the relationship between two categorical variables

Learning Objective

As a result of completing this investigation, students should be able to carry out a complete analysis of an experiment involving one or more categorical variables using counts, percents, and bar graphs to support their narrative conclusions.

Teaching Tips

We have designed this investigation so that students can formulate a plan for their experiment with little or no prompting, using only the first page of the student investigation to get started. Questions 1 through 7 then ask students to review their proposed design in light of several important issues before finalizing their experimental design in Question 8.

Obtain permission from your administration before allowing students to conduct the experiment. You may be required to get parental consent before students can participate in the experiment.

You will need to provide clear instructions to your students about obtaining informed consent, preserving anonymity and confidentiality, and ensuring subject's health and safety.

Next, students carry out their beverage preference experiment. Using the data they have collected, students are asked to perform an analysis and draw conclusions about students' preferences in Questions 9 through 11. Note that Question 10 focuses on the issue of whether order of presentation seems to have affected student preference, while Question 11 addresses the original research question.

Finally, students are asked to write a report about teenagers' preference for blue-colored beverages based on the results of this experiment. This question gives students a final opportunity to showcase their ability to analyze results from an experiment.

Suggested Answers to Questions

1. People may have a tendency to prefer the beverage they taste first (or last), regardless of the actual qualities of the beverages themselves (color, taste, etc.). That is, it is possible that the order in which people taste the beverages might affect their stated preference. If so, then you wouldn't want to present the same beverage first (or last) to more than about half of the subjects. Randomizing the order should help ensure that about half of the subjects taste one beverage first and about half taste the other beverage first. Then, any sizable differences that emerge in terms of preference for one beverage over the other should not be due to the order in which the beverages were presented.

2. One way to determine the order would be to flip a coin for each subject. If the coin shows "heads," then the subject would drink the clear beverage followed by the blue beverage. If the coin shows "tails," then the subject would drink the blue beverage followed by the clear beverage. Note that this method of randomly assigning the order could result in unequal numbers of subjects drinking the beverages in the two possible orders. An alternative method would be to put subjects' names on roughly identical slips of paper, drop them in a hat, and mix them up. Draw one slip at a time without looking. The person whose name is drawn first will drink clear then blue; the person whose name is drawn second would drink blue then clear, and so forth. Of course, you could use a modified version of the hat method by giving each subject a distinct numeric label, and then using a random number table or random number generator to select individuals one at a time. As before, the person whose name is drawn first will drink clear then blue; the person whose name is drawn second would drink blue then clear, and so forth.

3. The specific treatments in this experiment are "clear then blue" and "blue then clear."

4. One of the aims of the experiment is to see how color affects subjects' perceived preferences for a beverage. To study this, you must allow the subjects to see whether the beverage they are tasting is clear or blue. Hence, the subjects cannot be blind.

5. Answers will vary. Students should use some form of random selection to choose subjects to participate in the experiment. A true random sample of students may not be practical, but there's no need to go to the opposite extreme and use volunteers, either.

6. Answers will vary. If students use random selection to choose the subjects for their experiment, it should be reasonable to generalize the results to the larger population from which the subjects were selected. That's the benefit of random selection!

7. Answers will vary. One possible question is: "Of the two beverages that you tasted, which did you prefer?"

8. Answers will vary. Students' plans should include:

> Research question, clearly stated
>
> Subjects: how many; how they will be selected
>
> Explanatory variable and treatments
>
> How subjects will be assigned to treatment combinations
>
> Response variable: what will be measured and how

9. Answers will vary. Students should construct a well-labeled, comparative bar graph to display the categorical variable of drink preference for the two experimental groups. If the number of subjects in the two groups differs, then students should use percents rather than counts to compare subjects' drink preferences.

10. Answers will vary. Students should be evaluated based on the strength and clarity of the graphical evidence they provide about whether preference differs based on order of tasting.

11. Answers will vary. Students should be evaluated based on the strength and clarity of the graphical and numerical evidence they provide about whether students clearly prefer either blue or clear soda.

12. Answers will vary. Students should be evaluated based on the strength and clarity of the graphical and numerical evidence they provide in support of their recommendations.

Possible Extensions

Can people distinguish bottled water from tap water? Coke from Pepsi? Students could design and carry out a taste test experiment to help answer questions such as these.

After completing Section IV of the module, students could use simulation to test for a significant difference in preference to reinforce some of the ideas associated with statistical inference.

Investigation #12: Would You Drink Blue Soda? 

Corresponds to pp. 80-84
in Student Module

Does what you see affect your perception of how it tastes? If color can influence how people think a food tastes, what implications does this have for companies that make and market food and beverages?[1]

PepsiCo might be interested in your answer to these questions, as they have had two marketing failures based on introducing nontraditional colored beverages. In the early 1990s, PepsiCo introduced Pepsi Clear, a cola-flavored drink that was clear instead of brown in color. Pepsi Clear was later discontinued because sales were low. In 2002, PepsiCo tried again with Pepsi Blue.[2] Pepsi Blue was a berry-flavored cola drink that was blue in color. The Pepsi web site (*www.pepsi.com*) says that Pepsi Blue was "created by and for teens. Through nine months of research and development, Pepsi asked young consumers what they want most in a new cola. Their response: Make it berry and make it blue."

Unfortunately for PepsiCo, Pepsi Blue, like Pepsi Clear, was not a successful product, and it was discontinued a few years later. So what happened? Was the mistake adding a berry flavoring to cola, making the cola blue, or a combination of both?

In this investigation, you'll investigate whether teens have a preference for or a dislike for blue-colored soda.

Getting Started

To decide whether coloring a soda blue is a good or bad strategy if the drink is going to be marketed to teenagers, you will design and conduct an experiment, collect and analyze the data, and then make a recommendation.

For this experiment, you can start with a clear-colored soda, such as 7-Up or Sprite. Experiment with adding blue food coloring to the soda to create a "recipe" for a blue version of the soda. Food coloring is tasteless, so the addition of food coloring will not change the actual taste of the soda.

Once you have developed your new product, think carefully about how you would design an experiment to determine if teens have a preference for the clear soda or the blue soda.

Note: Be sure to discuss the ethical considerations involved in performing an experiment with human subjects. Your teacher will require you to obtain informed consent from all students (and possibly their parents) before they can participate in your experiment.

Once you have a plan in mind, answer the following questions. Be as specific as possible in your answers. It is OK to modify the design of your experiment if any of these

---

1   The page titled "Does the Color of Foods and Drinks Affect the Sense of Taste?" on the Neuroscience for Kids web site, *http://faculty.washington.edu/chudler/ coltaste.html*, has a list of references to studies that have examined how color affects perceived taste.

2   You can find an announcement describing the launch of Pepsi Blue at *http:// money.cnn.com/2002/05/07/news/companies/pepsi*.

questions reveal a weakness in your original plan. Now is the time to revise, before you actually carry out the experiment and collect the data!

1. In taste test experiments like the one you are designing, it is usual to randomize the order in which subjects taste the two drinks. That is, some subjects should taste the clear drink first and then the blue drink, while others should taste the blue drink first and then the clear. A random mechanism would be used to determine the order for each subject. Why do you think it is important to randomize the order in which the drinks are presented in an experiment of this type?

2. What would be a good way to determine the order (clear then blue or blue then clear) for each subject?

3. What are the two treatments for this experiment? *Hint:* In an experiment, subjects are assigned at random to one of the treatments.

4. Explain why it is not possible in this experiment to "blind" the subjects with respect to which experimental group they are in.

5. How will you select the subjects for your experiment, and how many subjects will participate? Be specific!

6. To what group, if any, will you be able to generalize the results of your experiment? Explain why you think it is reasonable to generalize to this particular group.

7. What question will you ask each subject after he or she has tasted the two sodas? Make sure that you will be able to determine from the response which of the two drinks was preferred.

8. After considering your answers to questions 1 through 7 and modifying your plan as needed, write a summary of your plan for conducting the experiment on *separate paper*. Include enough detail that someone who has not been part of your design team could read the summary and be able to carry out the experiment as you intended. Be sure to address ethical issues of using human subjects.

After your teacher has approved your experimental plan, carry out the experiment and collect data. Be sure to record the order in which the two drinks were tasted and the response for each subject.

Once you have collected the data, use it to fill in the four cells of the table below.

		Order	
		Clear then Blue	Blue then Clear
Preference	Clear		
	Blue		

9. Construct a graphical display that allows you to compare the preferences for the two experimental groups (clear then blue and blue then clear).

10. Based on your display, do you think there is a difference in preference for the two experimental groups? That is, do you think the order in which the drinks were tasted makes a difference? Explain.

11. Based on the data from this experiment, do you think there is a preference for one of the drinks (clear or blue) over the other? Explain, justifying your answer using the data from the experiment.

12. Write a report that makes recommendations to a soft drink company that is considering introducing a blue soft drink that will be marketed to teens. Include appropriate data and graphs to support your recommendations.

Teacher Notes for Section IV: Drawing Conclusions

In the investigations of this section, students are given a brief introduction to the ideas of statistical inference—the process of drawing conclusions about a population using data from a sample. The three investigations in this section illustrate how sample data can be used to estimate a population characteristic and evaluate the plausibility of a claim about a population. Three important ideas of statistical inference are introduced—margin of error, convincing evidence, and significant difference.

The three investigations in this section are:

Investigation #13: The Internet—Information or Social Highway?

Students explore the role of sampling variability in the context of estimating a population proportion and are introduced to the idea of a margin of error.

Investigation #14: Evaluating the MySpace Claim

Students explore the meaning of "convincing evidence" in the context of using sample data to evaluate the plausibility of a claim about a population proportion.

Investigation #15: Are Teens the Same Everywhere?

Students investigate the meaning of "significant difference" in the context of using sample data to determine if there is a significant difference between the proportions of students at two schools who respond in a particular way to a survey question.

Prerequisites

Students should be able to:

Design and implement a reasonable sampling plan in order to collect survey data

Explain why the value of a sample statistic, such as a sample proportion, varies from sample to sample

Summarize numerical data using a dotplot

Learning Objectives

As a result of completing this section, students should be able to:

Understand what is meant by a conservative margin of error in the context of estimating a population proportion

Compute a conservative margin of error when a proportion from a random sample is used to estimate a population proportion

Understand how characterizing sampling variability enables us to determine what constitutes "convincing evidence"

Use simulation to determine what values of an observed sample proportion would be considered unusual (not likely to have resulted due solely to sampling variability) for a given population proportion

Use data from a random sample to draw a conclusion about a population proportion

Understand what is meant by the statement that there is a "significant difference" between two groups

Use simulation to determine what values of the observed difference in sample proportions would be unlikely to occur if there is no difference in the corresponding population proportions

Formulate a question about the difference between two population proportions and then use simulation and sample data to provide an answer to the question posed

Teaching Tips

Understanding and characterizing sampling variability, the sample-to-sample differences that occur just due to random selection, is the key to understanding the logic of statistical inference. Spend time talking about sampling variability in each of the investigations of this section.

It is worth spending some time talking about the examples in the Section IV Overview prior to beginning the investigations of this section. This will help students see where the investigations are leading them.

If you have a small class, you may need to have each student construct more than five simulated samples in the simulations that appear in the investigations of this section. Try to have at least 100 simulated samples when the class information is combined (more is even better).

THE OBJECTIVE OF MANY STATISTICAL STUDIES IS TO USE SAMPLE DATA TO TELL US something about the population from which the sample was selected. As we have seen in the previous sections, if the sample is selected in an appropriate way and involves some type of random selection, it may be reasonable to regard the sample as representative of the population of interest. Suppose that 15 in a sample of 50 randomly selected students favor banning soda machines from campus—a proportion of 15/50 = .30 or, equivalently, 30%. Is it also reasonable to say that exactly 30% of the students at your school favor banning soda machines from campus? The value of the sample proportion depends on the outcome of the random selection—which 50 students are selected to participate. As a consequence, the value of the sample proportion will vary from one random sample to another, and we can't expect that the sample proportion will be exactly equal to the actual value of the population proportion.

Does this mean the sample proportion doesn't tell us anything about the population proportion? Fortunately, the answer to that question is no! If the sample has been chosen appropriately and can be regarded as representative of the population, we can expect that the sample value will, on average, be "close" to the true population value. But, to really tell us something useful about the population, we need to be a bit more specific about what we mean by "close."

In this section, we will begin to investigate how we can use data from a well-designed study to draw conclusions about a population. This requires an understanding of the nature of **sampling variability**—the chance differences that occur from one random sample to another as a consequence of random selection.

Statistical inference is the process of drawing conclusions about a population using data from a sample. In most statistical studies, sample data is collected so that the investigator can either estimate some population characteristic of interest (such as the proportion of students at your school who favor the ban of soda machines on campus) or evaluate the plausibility of some claim that has been made about the population (for example, a claim that more than 50% of the students at the school favor a ban on soda machines).

In an **estimation** situation, we need to understand sampling variability to be able to assess how close an estimate from a sample is likely to be to the actual value of the corresponding population characteristic. In published reports, you will often see statements that include a **margin of error**. For example, the analysis of data from a survey on public support for the president might include a statement such as "The proportion of U.S. adults who believe the president is doing a good job is .45 (45%) with a margin of error of .03 (3%)." The reported margin of error acknowledges that the true population proportion is not likely to be *exactly* .45 and indicates that, based on what was seen in the sample, plausible values for the population proportion might be anything between .42 and .48. How is the margin of error computed? It is based on an assessment of sampling variability. Investigation 13 illustrates the reasoning that leads to a margin of error calculation.

The second way data from a sample is typically used to draw a conclusion about a population is in the evaluation of the plausibility of a claim about the population. For example, the *Youth Monitor* report titled "Coming of Age in America: Part IV – The MySpace Generation" (*www.greenbergresearch.com*) includes the following statement:

> Half (52%) of 18–24 year olds report they have a page on MySpace. A third report membership on Facebook, though that number rises to over half (54%) among students.

What about students at your school? Does a majority (more than 50%) have a page on MySpace? Suppose the proportion in a random sample of students from your school who report that they have a MySpace page was .52. Can you conclude that a majority of *all* students at the school have a MySpace page? This requires some careful thought. There are two reasons why the sample proportion might have been larger than .50. One reason is sampling variability—we don't expect the sample proportion to be exactly equal to the population proportion. So, maybe the population proportion is .50 (or maybe even something smaller) and a sample proportion of .52 is "explainable" just due to sampling variability. If this is the case, we can't interpret the sample proportion of .52 as convincing evidence that a majority of *all* students at your school have a MySpace page. However, another reason the sample proportion might have been greater than .50 is that the true population proportion is, in fact, greater than .50. Is .52 enough larger than .50 that the difference can't be explained as being due to sampling variability alone? If so, we would say that, based on the sample data, there is convincing evidence that more than half of the students at the school have a MySpace page. How do we make this determination? Again, it is based on an assessment of sampling variability. Investigations 14 and 15 illustrate the reasoning that enables us to use data from a random sample to evaluate the plausibility of a claim about a population.

The investigations in this section are designed to introduce you to reasoning about sampling variability in each of these two types of settings—estimation and evaluating the plausibility of a claim. In this brief introduction, we can illustrate the reasoning, but for a more complete treatment of the inferential process, we encourage you to consider taking a course devoted to statistics and data analysis!

Teacher Notes for Investigation 13:
The Internet—Information or Social Highway?

This investigation illustrates how sampling variability can be characterized so that a margin of error can be associated with an estimate based on sample data. After exploring sampling variability in the context of estimating a population proportion, students carry out a simple survey. They then use the resulting data to estimate the proportion of students that agree with a particular statement. Finally they compute a conservative margin of error for the estimate.

Prerequisites

Students should be able to:

> Understand how a margin of error (introduced in the section Overview) is interpreted.

> Design and implement a reasonable sampling plan in order to collect survey data

Learning Objectives

As a result of completing this investigation, students should be able to:

> Describe what is meant by a conservative margin of error in the context of estimating a population proportion

> Compute a conservative margin of error when a proportion from a random sample is used to estimate a population proportion

Teaching Tips

Spend some time making sure students understand what a conservative margin of error is. Thoroughly discuss the definition in the investigation introduction.

Be sure to work through the simulation example in the investigation introduction with the class prior to having students begin to answer the questions of the investigation.

If you have a small class, you may need to have each student construct more than five simulated samples in response to question 2. Try to have at least 100 simulated samples when the class information is combined in question 3 (more is even better).

If you have time in class, you might let students share what they have written in response to question 19. Another possibility is to have students pair up and critique each other's answers.

Suggested Answers to Questions

1. Answers will vary. For example, for the simulated sample of the investigation introduction, the number of successes in the sample was 27 and the proportion of successes for the sample was .675.

2. Answers will vary. Most simulated proportions will be between .52 and .68, although you may occasionally see values that are outside of this range.

3. Answers will vary. As an example, one set of 100 simulated samples of size 40 resulted in the frequencies shown in the table below.

Sample Proportion	Tally	Frequency	Sample Proportion	Tally	Frequency
0.000		0	0.525		11
0.025		0	0.550		9
0.050		0	0.575		9
0.075		0	0.600		12
0.100		0	0.625		11
0.125		0	0.650		11
0.150		0	0.675		11
0.175		0	0.700		5
0.200		0	0.725		3
0.225		0	0.750		1
0.250		0	0.775		0
0.275		0	0.800		3
0.300		0	0.825		0
0.325		0	0.850		0
0.350		0	0.875		0
0.375		0	0.900		0
0.400		1	0.925		0
0.425		2	0.950		0
0.450		3	0.975		0
0.475		3	1.000		0
0.500		5			

4. Answers will vary. For the example simulated results given in the table on the previous page, the answers would be as follows:

(a) .450

(b) .750

(c) .450, .750

5. Answers will vary. For the example simulated results given in the table on the previous page, the endpoints of the interval in question 4(c) are both .150 away from .6. So a reasonable answer here would be that about 95% of the time, the sample proportion was within .150 of the actual population proportion of .6.

6. Conservative margin of error = .158.

7. Answers will vary. For the example simulated results given in the table on the previous page, the conservative margin of error is larger, which is expected since it is a conservative estimate. Note that occasionally a simulation will produce an answer that is larger than the conservative estimate. If this happens, it is probably because there were not a sufficient number of trials in the simulation to get good approximations for the answers to question 4.

8. Any assignment that uses the digits from 0 to 9 and assigns two of them to represent a success will work. For example, letting 0 and 1 represent a success and 3, 4, ..., 9 represent a failure is one possibility. There are also possible correct answers that begin with something other than the digits from 0 to 9.

9. Answers will vary.

10. Answers will vary.

11. Answers will vary. As an example, one set of 100 simulated samples of size 40 resulted in the frequencies below.

Sample Proportion	Tally	Frequency	Sample Proportion	Tally	Frequency
0.000		0	0.525		0
0.025		0	0.550		0
0.050		0	0.575		0
0.075		2	0.600		0
0.100		5	0.625		0
0.125		9	0.650		0
0.150		14	0.675		0
0.175		15	0.700		0
0.200		13	0.725		0
0.225		14	0.750		0
0.250		9	0.775		0
0.275		10	0.800		0
0.300		3	0.825		0
0.325		4	0.850		0
0.350		1	0.875		0
0.375		1	0.900		0
0.400		0	0.925		0
0.425		0	0.950		0
0.450		0	0.975		0
0.475		0	1.000		0
0.500		0			

12. The observed sample proportions tend to center around .6 for a population with 60% successes, whereas the observed proportions tend to center around .2 for a population with 20% successes. In addition, the observed sample proportions from the population with 20% successes tend to cluster more tightly around .2 (there is less sample-to-sample variability) than the observed sample proportions from the population with 60% successes tend to cluster around .6.

13. Answers will vary. For the example simulated results given in the table on the previous page, the answers would be as follows:

(a) .100

(b) .325

(c) .100, .325

14. Answers will vary. For the example simulated results given in the table on the previous page, the endpoints of the interval in question 13(c) are .100 and .125 away from .2. So a reasonable answer here would be that about 95% of the time, the sample proportion was within .125 of the actual proportion of .2.

15. Answers will vary. For the example simulated results given on the previous page, the conservative margin of error is larger, which is expected since it is a conservative estimate. Note that occasionally a simulation will produce an answer that is larger than the conservative estimate. If this happens, it is probably because there were not a sufficient number of trials in the simulation to get good approximations for the answers to question 13.

16. Yes. The estimated margins of error when $n = 40$ and the population has 60% successes is about ___ (answer from question 5; for our example, this was .150). When $n = 40$ and the population has 20% successes, the estimated margin of error is about ___ (answer from question 14; for our example, this was .125). Both are smaller than the conservative margin of error for a sample size of 40, which is .158 (from question 6). Note that occasionally a simulation will produce an answer that is larger than the conservative estimate. If this happens, it is probably because there was not a sufficient number of trials in the simulation to get good approximations.

17. Answers will vary. Ideal responses should incorporate random selection or include a convincing justification explaining why the selected method is likely to produce a sample that is representative of the population.

18. Answers will vary.

19. Answers will vary.

Investigation #13: The Internet —Information or Social Highway?

The report "Coming of Age in America: Part IV—The MySpace Generation" referenced in the Overview to this section also included the following:

> For many young people, the Internet represents not just an information super-highway but, indeed, a social highway. In fact, nearly two thirds (64%) agree that "I don't know how I would keep up with my friends or family if I didn't have the Internet."

Corresponds to pp. 87-93
in Student Module

What proportion of the students at your school agree with the statement above? In this investigation, you will carry out a simple survey and then use the resulting data to estimate the proportion who agree. You will also see how we can obtain a conservative margin of error for your estimate.

Let's begin by taking a look at margin of error. A **conservative estimate of the margin of error** when a proportion from a random sample is used as an estimate of the corresponding population proportion is $\frac{1}{\sqrt{n}}$, where n denotes the sample size. For a sample size of 40, it is reasonable to use $\frac{1}{\sqrt{n}}$ as a conservative estimate of margin of error for sample proportions between .15 and .85. For larger sample sizes, this way of computing a conservative margin of error is reasonable for an even larger range of values of the sample proportion. When we say that this is a conservative estimate of the margin of error, we mean that the actual margin of error will be equal to or smaller than this conservative value. That is, when we use this conservative estimate, we may be overstating our potential error. This is generally considered to be a better alternative than understating potential error. It is possible to obtain more precise estimates of the margin of error, but that is beyond the scope of this introductory investigation.

To see why it is reasonable to use $\frac{1}{\sqrt{n}}$ as a conservative estimate of the margin of error, we need to consider the role of sampling variability and how the sample proportion varies from one random sample to another. To do this, we will carry out a simulation.

How much will sample proportions vary from one random sample to another and how much will these sample proportions tend to differ from the actual value of the corresponding population proportion? Suppose that the proportion of students at your school who have some characteristic of interest is actually .60. Think of this as a population where 60% have the characteristic we are interested in (called them "successes"). What kind of sample proportions would you expect to see for random samples of size 40 from this population? We can find out by creating a hypothetical population with 60% "successes." One way to create such a population is to use random digits. Because each of the digits 0, 1, 2, …, 9 is equally likely to occur in a list of random digits, if we think of the digits 1, 2, 3, 4, 5, and 6 as representing individuals who are successes and the digits 0, 7, 8, and 9 as representing individuals who are not successes, we can view a collection of random digits as a population with 60% successes.

151

We can now take a random sample of size 40 from this hypothetical population by selecting 40 random digits. Each member of the sample (each digit) can then be classified according to whether it is a success, and then the sample proportion of success can be computed. For example, consider the list of 40 random digits below.

$$0\ \underline{5}\ \underline{2}\ \underline{3}\ \underline{5}\ 7\ \underline{2}\ \underline{3}\ \underline{2}\ \underline{3}\ 0\ 7\ \underline{1}\ \underline{1}\ 0\ 0\ \underline{3}\ \underline{2}\ \underline{5}\ 8\ \underline{1}\ \underline{2}\ \underline{4}\ \underline{6}\ 9\ \underline{4}\ \underline{5}\ \underline{3}\ \underline{1}\ 7\ 0\ \underline{2}\ \underline{1}\ \underline{5}\ 9\ \underline{3}\ 9\ \underline{4}\ 7\ \underline{6}$$

Identifying the 1s, 2s, 3s, 4s, 5s, and 6s as successes (the underlined digits above), we could then compute the proportion of successes for this random sample:

$$\text{sample proportion} = \frac{\text{number of successes in the sample}}{n} = \frac{27}{40} = .675$$

Repeating this process a large number of times and noting the values of the sample proportions obtained allows us to see what values of the sample proportion are expected as a result of sampling variability alone.

1. Use a sequence of random digits to represent a random sample from a population that has 60% successes (a population proportion of .60). Identifying successes as described in the paragraph above, determine the number of successes in the sample and the sample proportion.

Number of successes in the sample =

Proportion of successes for this sample =

2. Record the sample proportion obtained in step 1 in the table below. Then, repeat the process four more times. Enter each of the four new sample proportions into the table below.

Sample Number	1	2	3	4	5
Sample Proportion					

3. Enter a tally mark for each of your observed sample proportions in the table on the following page. Then enter a tally mark for each of the five sample proportions observed by other students in your class. You can do this by making tally marks as each student reads his or her sample proportions. When finished, count the tally marks for each row in the table and enter the total in the frequency column for the appropriate row.

4. Use the information in the table from step 3 to complete the following sentences.

(a) About 2.5% of the observed sample proportions are less than _______.

(b) About 2.5% of the observed sample proportions are greater than _______.

(c) About 95% of the observed sample proportions are between _______ and _______.

Sample Proportion	Tally	Frequency	Sample Proportion	Tally	Frequency
0.000			0.525		
0.025			0.550		
0.050			0.575		
0.075			0.600		
0.100			0.625		
0.125			0.650		
0.150			0.675		
0.175			0.700		
0.200			0.725		
0.225			0.750		
0.250			0.775		
0.275			0.800		
0.300			0.825		
0.325			0.850		
0.350			0.875		
0.375			0.900		
0.400			0.925		
0.425			0.950		
0.450			0.975		
0.475			1.000		
0.500					

5. Now try to complete the following statement:

> About 95% of the time, the sample proportion was within _______ of the actual population proportion (.60).

Hint: Look at your answer to part (c) of question 4. How far are the endpoints of the interval you give there from .60?

6. Compute the conservative margin of error for a sample size of 40.

Conservative margin of error $= \dfrac{1}{\sqrt{n}} =$

7. How does the conservative margin of error from step 6 compare to your number from step 5?

What if the proportion of successes in the population was .2 instead of .6? What kind of sample proportions should we expect to observe? We can investigate this by carrying out a simulation that is similar to the one just carried out. Only a few modifications are needed.

8. If a sequence of random digits is to be used to represent a random sample from a population that has 20% successes, what digits will you use to represent successes in the sample?

9. Use a sequence of random digits to represent a random sample of size 40 from a population that has 20% successes (a population proportion of .20). Determine the number of successes in the sample and the sample proportion.

Number of successes in the sample =

Proportion of successes for this sample =

10. Record the sample proportion obtained in step 9 in the table below. Then repeat the process four more times. Enter each of the four new sample proportions into the table.

Sample Number	1	2	3	4	5
Sample Proportion					

11. Enter a tally mark for each of your observed sample proportions in the table on the next page. Then enter a tally mark for each of the five sample proportions observed by other students in your class. You can do this by making tally marks as each student reads his or her sample proportions. When finished, count the tally marks for each row in the table and enter the total in the frequency column for the appropriate row.

12. Write a few sentences describing how the observed sample proportions for random samples of size 40 differ for a population with 60% successes and a population with 20% successes.

Sample Proportion	Tally	Frequency	Sample Proportion	Tally	Frequency
0.000			0.525		
0.025			0.550		
0.050			0.575		
0.075			0.600		
0.100			0.625		
0.125			0.650		
0.150			0.675		
0.175			0.700		
0.200			0.725		
0.225			0.750		
0.250			0.775		
0.275			0.800		
0.300			0.825		
0.325			0.850		
0.350			0.875		
0.375			0.900		
0.400			0.925		
0.425			0.950		
0.450			0.975		
0.475			1.000		
0.500					

13. Use the information in the table to complete the following sentences.

(a) About 2.5% of the observed sample proportions are smaller than ________.

(b) About 2.5% of the observed sample proportions are greater than ________.

(c) About 95% of the observed sample proportions are between ________ and ________.

14. Now try to complete the following statement:

About 95% of the time, the sample proportion was within ________ of the actual population proportion (.20).

Hint: Look at your answer to part (c) of question 13. How far are the endpoints of the interval you give there from .20?

15. How does the conservative margin of error for a sample of size 40 (see step 6) compare to your number from step 14?

16. Consider the following statement from the introduction to this investigation:

> When we use this conservative estimate, we may be overstating our potential error. This is generally considered to be a better alternative than understating potential error. It is also worth noting that the conservative estimate overstates to a larger degree the more that the sample proportion differs from .5.

Is this statement supported by the two simulations that you have carried out? Explain.

Now that you understand what is meant by conservative margin of error and know how to compute it, you are ready to see how this is used when reporting survey data. Recall the question posed at the beginning of this investigation:

> What proportion of the students at your school agree with the statement "I don't know how I would keep up with my friends or family if I didn't have the Internet"?

17. As a class, discuss how you will go about selecting a sample of students at your school to participate in a survey. Remember that the sample should be selected in such a way that it will be reasonable to generalize from the sample to the population of all students at the school. Be sure to consider the ethical issues involved in conducting a survey—informed consent, right of refusal, and preserving anonymity and confidentiality. Write a brief summary of your plan for carrying out the survey.

18. Once your teacher has approved your plan, carry out the sampling that is required. Then, ask each individual selected for inclusion in the sample whether he or she agrees with the statement "I don't know how I would keep up with my friends or family if I didn't have the Internet." Use the resulting data to compute the sample proportion and a conservative margin of error.

Sample size =

Number in sample who agree with statement =

Sample proportion =

Conservative margin of error =

19. Write a paragraph describing the results of your survey. Be sure to reference the estimate of the proportion of students at your school who agree with the statement and the margin of error. Also, indicate whether you think the statement from the report referenced at the beginning of this investigation is accurate for your school and why or why not. (The report said nearly two-thirds of young people agree with the statement.)

Teacher Notes for Investigation 14: Evaluating the MySpace Claim

IN THIS INVESTIGATION STUDENTS ARE INTRODUCED TO ONE OF THE MOST IMPORTANT ideas of statistical inference—what constitutes "convincing evidence" against a claim about a population characteristic. Students carry out a survey of students at their school and then use the resulting data to determine if there is convincing evidence that a majority of students at the school have a MySpace page.

Prerequisites

Students should be able to:

Explain why the value of a sample statistic, such as a sample proportion, varies from sample to sample

Design and implement a reasonable sampling plan to collect survey data

Learning Objectives

As a result of completing this investigation, students should be able to:

Describe how characterizing sampling variability enables us to determine what constitutes "convincing evidence"

Use simulation to determine what values of an observed sample proportion would be considered unusual (not likely to have resulted due solely to sampling variability) for a given population proportion

Use data from a random sample to draw a conclusion about a population proportion

Teaching Tips

Be sure to work through the simulation example in the investigation introduction with the class prior to having students begin to answer the questions of the investigation.

If you have a small class, you may need to have each student construct more than five simulated samples in response to question 2. Try to have at least 100 simulated samples when the class information is combined in question 3 (more is even better).

If you have time in class, you might let students share what they have written in response to question 10. Another possibility is to have students pair up and critique each other's answers.

Suggested Answers to Questions

1. Answers will vary, but they should look something like the example in the introduction to the investigation.

2. Answers will vary.

3. Answers will vary. As an example, one set of 100 simulated samples of size 60 resulted in the frequencies on the following page.

Sample Proportion	Frequency	Sample Proportion	Frequency
0.017	0	0.517	12
0.033	0	0.533	8
0.050	0	0.550	13
0.067	0	0.567	7
0.083	0	0.583	3
0.100	0	0.600	3
0.117	0	0.617	4
0.133	0	0.633	2
0.150	0	0.650	2
0.167	0	0.667	0
0.183	0	0.683	0
0.200	0	0.700	0
0.217	0	0.717	0
0.233	0	0.733	0
0.250	0	0.750	0
0.267	0	0.767	0
0.283	0	0.783	0
0.300	0	0.800	0
0.317	0	0.817	0
0.333	0	0.833	0
0.350	0	0.850	0
0.367	1	0.867	0
0.383	2	0.883	0
0.400	3	0.900	0
0.417	2	0.917	0
0.433	5	0.933	0
0.450	7	0.950	0
0.467	5	0.967	0
0.483	11	0.983	0
0.500	10	1.000	0

4. Answers will vary. For the example simulated results given in the table above, the answer would be:

About 5% of the observed sample proportions are greater than .617.

5. Answers will vary. For the example simulated results given in the table on the previous page, the answer would be:

> If we take a random sample of 60 students from our school and we find that the proportion who have a MySpace page is equal to or greater than .617, we will conclude that there is convincing evidence that a majority of students at the school have a MySpace page.

6. Based on the simulation, we can see that when the population proportion is .5 and the sample size is 60, it is relatively rare to observe a sample proportion as large as the value identified in question 5 just due to sampling variability. By relatively rare, we mean that this would happen less than 5% of the time *just due to sampling variability*.

7. Answers will vary. Be sure the final sampling plan involves some type of random selection.

8. Answers will vary.

9. Answers will vary. If the observed sample proportion from question 8 is less than the value identified in question 5, conclusion 2 should be chosen. If the observed sample proportion from question 8 is equal to or greater than the value identified in question 5, conclusion 1 should be chosen.

10. Answers will vary.

 # Investigation #14: Evaluating the MySpace Claim

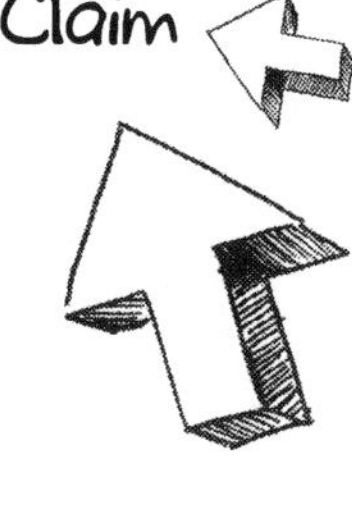

In the Section IV Overview, the following question was posed: Does a majority (more than 50%) of students at your school have a page on MySpace? In this investigation, you will carry out a survey of 60 students at your school. Then, you will use the resulting data to determine if there is convincing evidence that a majority of students at your school have a MySpace page.

Before we collect data, let's consider what it would take to convince us that a majority of students at the school have a MySpace page. Suppose we select a random sample and find that the proportion in the sample who report they have a MySpace page is less than .50 (less than 50%). That certainly wouldn't be convincing evidence that more than 50% have such a page. But what if our sample proportion is greater than .50? Should we be "convinced" that a majority of *all* students (not just those in the sample) have a MySpace page? This decision isn't so easy!

What we need to see to be *convinced* is not just a sample proportion greater than .50—it has to be *enough* greater that it is not likely to have occurred just by chance due to sampling variability. That is, to be convinced, we would need to rule out sampling variability—those chance differences that occur from one sample to another as a consequence of random selection—as a plausible explanation for what we see in the sample. For our MySpace example, if we decide that the sample proportion is enough larger than .50 that we don't think it is just due to sampling variability, we would conclude that there is convincing evidence that the actual proportion in the population is larger than .50.

So, how do we decide just how much greater than .50 our sample proportion must be in order to convince us that what we are seeing is not just sampling variability? To figure this out, we need to know what kind of sample proportions are likely to be observed just by chance when the population proportion is .50 or less (not a majority). If we see one of these proportions, we would not want to say we were convinced that a majority of students have a MySpace page.

The first steps in this investigation use simulation to explore what values of the sample proportion would *not* be convincing evidence of a majority. Suppose that in fact 50% or fewer of the students at your school have a MySpace page. We will focus on the most extreme of these possible population values—50%. We use 50% (a proportion of .50) in the simulation that follows because to be convinced that a majority of students have a MySpace page, we need to be convinced that the population proportion is greater than .50.

To determine what sample proportions are likely to be observed when the population proportion is .50, we will use a hypothetical population of random digits. We create a population with 50% "successes" by designating the digits 1, 2, 3, 4, and 5 to represent individuals who are successes and the digits 6, 7, 8, 9, and 0 to represent individuals who are not successes.

Corresponds to pp. 94-98 in Student Module

We can now take a random sample of size 60 from this hypothetical population by selecting 60 random digits. Each member of the sample (each digit) can then be classified according to whether it is a success. Then, the sample proportion of successes can be computed. For example, consider the list of 60 random digits below.

8 4 6 6 4 3 7 8 6 7 3 0 1 3 5 4 6 7 6 2 6 3 8 2 9 9 1 9 0 0
3 0 6 3 7 7 0 5 9 1 8 1 3 4 3 6 5 4 0 3 0 9 2 2 5 9 7 1 5 9

Identifying the 1s, 2s, 3s, 4s, and 5s as successes (the underlined digits above), we could then compute the proportion of successes for this random sample:

$$\text{sample proportion} = \frac{\text{number of successes in the sample}}{n} = \frac{28}{60} = .467$$

Repeating this process a large number of times and noting the values of the sample proportions obtained allows us to see what values of the sample proportion are expected as a result of sampling variability alone.

1. Use a sequence of 60 random digits to represent a random sample of size 60 from a population that has 50% successes (a population proportion of .50). Identifying successes as described in the paragraph above, determine the number of successes in the sample and the sample proportion.

Number of successes in the sample =

Proportion of successes for this sample =

2. Record the sample proportion obtained in step 1 in the table below. Then, repeat the process four more times. Enter each of the four new sample proportions into the table.

Sample Number	1	2	3	4	5
Sample Proportion					

3. Enter a tally mark for each of your observed sample proportions in the table on the next page. Then, enter a tally mark for each of the five sample proportions observed by other students in your class. You can do this by making tally marks as each student reads his or her sample proportions. When finished, count the tally marks for each row in the table and enter the total in the frequency column for the appropriate row.

4. Use the information in the table from step 3 to complete the following sentence.

About 5% of the observed sample proportions are greater than _______.

5. Based on the statement in step 4, it is reasonable to say that when the true population proportion is .50, a value as large as or larger than the number in step 4 would

Sample Proportion	Tally	Frequency	Sample Proportion	Tally	Frequency
0.017			0.517		
0.033			0.533		
0.050			0.550		
0.067			0.567		
0.083			0.583		
0.100			0.600		
0.117			0.617		
0.133			0.633		
0.150			0.650		
0.167			0.667		
0.183			0.683		
0.200			0.700		
0.217			0.717		
0.233			0.733		
0.250			0.750		
0.267			0.767		
0.283			0.783		
0.300			0.800		
0.317			0.817		
0.333			0.833		
0.350			0.850		
0.367			0.867		
0.383			0.883		
0.400			0.900		
0.417			0.917		
0.433			0.933		
0.450			0.950		
0.467			0.967		
0.483			0.983		
0.500			1.000		

occur only about 5% of the time just by chance. That means that it would be fairly unusual to see a value this large if the population proportion is .50 (and even less likely if the population proportion were something smaller than .50). Use the number from step 4 to complete the following decision statement:

> If we take a random sample of 60 students from our school and we find that the proportion who have a MySpace page is equal to or greater than _______, we will conclude that there is convincing evidence that a majority of students at the school have a MySpace page.

6. Write a few sentences explaining why the decision statement in step 5 makes sense.

7. We are almost ready to collect some data. As a class, discuss how you will go about conducting a survey involving 60 students at your school. Remember that the sample should be selected in such a way that it will be reasonable to generalize from the sample to the population of all students at the school. Be sure to consider the ethical issues involved in conducting a survey—informed consent, right of refusal, and preserving anonymity and confidentiality. Write a brief summary of your plan for carrying out the survey.

8. Implement the sampling plan described in the previous step, asking each chosen individual whether he or she has a MySpace page. Use the resulting data to compute the sample proportion.

Sample size =

Number in sample who agree with statement =

Sample proportion =

9. Based on the observed sample proportion and the decision statement from step 5, which of the following possible conclusions is appropriate?

Conclusion 1: There is convincing evidence that a majority of students at our school have a MySpace page.

Conclusion 2: There is not convincing evidence that a majority of students at our school have a MySpace page.

10. Write a paragraph describing your conclusions. Include an assessment of whether it is plausible that 50% or fewer of the students at your school have a MySpace page and that what was observed in the sample can be explained by sampling variability alone.

Teacher Notes for Investigation 15: Are Teens the Same Everywhere?

This investigation introduces students to another important idea in statistical inference—what it means when we say there is a "significant difference" between two groups. Data collected from students at two schools, one in the United States and one in Sweden, are used to decide if there is a significant difference between the two schools in the proportion of students who report that they would probably get home on time when given a curfew. Students then formulate and answer other questions that can be answered on the basis of the survey data provided.

Prerequisites

Students should be able to:

Explain why the value of a sample statistic, such as a sample proportion, varies from sample to sample

Summarize numerical data using a dotplot

Learning Objectives

As a result of completing this investigation, students should be able to:

Understand what is meant by the statement that there is a "significant difference" between two groups

Use simulation to determine what values of the observed difference in sample proportions would be unlikely to occur if there is no difference in the corresponding population proportions

Formulate a question about the difference between two population proportions and then use simulation and sample data to provide an answer to the question posed

Teaching Tips

If you have a small class, you may need to have each student construct more than five simulated samples in response to question 5. Try to have at least 100 simulated samples when the class information is combined in question 6 (more is even better).

If you have time in class, you might let students share what they have written in response to question 9. Another possibility is to have students pair up and critique each other's answers.

In question 10, teams of size four or five work well. Encourage students to formulate a question that is of interest to them.

It might be a good idea to have each of the teams formed for question 10 explain how they will carry out their simulation and get approval prior to actually beginning the simulation. This will ensure that students don't spend a lot of time carrying out a simulation that won't provide the information needed in order for them to proceed to a conclusion.

Be sure that each team carrying out a simulation for question 10 has at least 100 simulated samples before using the simulation results to answer the question posed.

If you have time in class, have each team present the question they chose and their conclusion. Ask each team to provide a justification for the conclusion.

Suggested Answers to Questions

1. See the table below.

	Home on Time	Not Home on Time	Total
Milwaukee Sample	19	41	**60**
Sweden Sample	14	42	**56**
Total	**33**	**83**	**116**

2. 14/56 = .25

3. Answers will vary. One example is given in the investigation:

For example, if the number in the highlighted cell was 16, the resulting table, proportions, and difference would be those in the table below.

	Home on Time	Not Home on Time	Total
Milwaukee Sample	16	44	**60**
Sweden Sample	17	39	**56**
Total	**33**	**83**	**116**

Home on Time proportion of Milwaukee sample = 16/60 = .27

Home on Time proportion of Sweden sample = 17/56 = .30

Difference in sample proportions (Milwaukee − Sweden) = .27 − .30 = −.03

4. Answers will vary.

5. Answers will vary.

6. Answers will vary. As an example, one set of 100 simulated differences resulted in the data at the top of the following page.

```
 0.02   -0.19   -0.12    0.02   -0.02   -0.19    0.02    0.12   -0.09
 0.05   -0.12    0.12   -0.02   -0.02   -0.05    0.05    0.05   -0.12
-0.09    0.05   -0.02   -0.02   -0.02   -0.02   -0.09    0.05   -0.09
-0.05   -0.02   -0.09    0.02   -0.16   -0.16   -0.05   -0.05    0.02
 0.15    0.15    0.05   -0.09    0.05    0.02    0.08    0.02   -0.05
 0.22    0.02    0.02   -0.02    0.08   -0.05    0.05   -0.09    0.08
 0.05    0.08    0.08   -0.09   -0.02   -0.05    0.05   -0.05    0.12
-0.09   -0.02    0.08   -0.05    0.12   -0.02   -0.09    0.05    0.05
-0.12    0.08   -0.12   -0.02   -0.02    0.02    0.08    0.05    0.05
-0.05    0.12    0.05    0.08   -0.05   -0.05   -0.09    0.19   -0.09
-0.12   -0.02   -0.05   -0.02    0.08    0.08    0.05    0.02   -0.09
 0.05
```

7. Answers will vary. As an example, using the set of 100 simulated differences above resulted in the following:

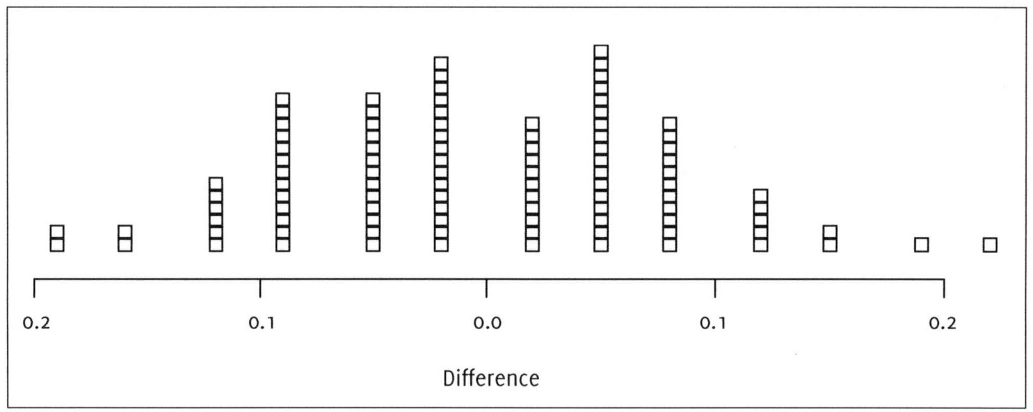

8. Answers will vary. As an example, using the set of 100 simulated differences above would result in the following:

> Based on the dotplot, an observed difference in sample proportions of .07 would not be unusual, even when there is no difference in the population proportions. Twenty of the 100 simulated samples produced a difference in sample proportions of .07 or larger. A difference in sample proportions of .07 is the kind of difference that would be expected just due to sampling variability alone.

9. Answers will vary. As an example, using the set of 100 simulated differences above would result in the following:

> A difference in sample proportions of .07 is the kind of difference that would be expected just due to sampling variability alone. Therefore, there is not a significant difference in the proportion for the two schools.

10. Answers will vary depending on the questions chosen.

# Investigation #15: Are Teens the Same Everywhere?

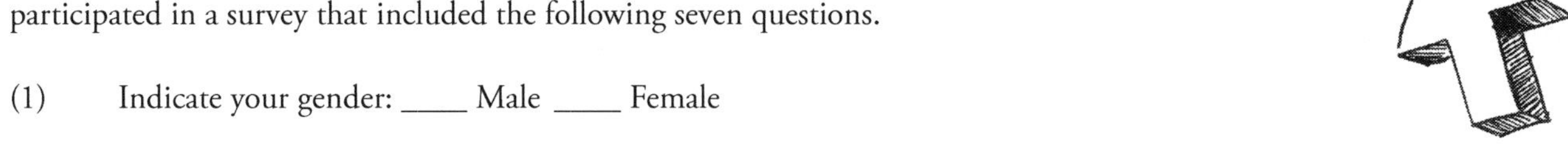

Students from a high school in Milwaukee, Wisconsin, and from a high school in Sweden participated in a survey that included the following seven questions.

(1)   Indicate your gender: _____ Male _____ Female

Corresponds to pp. 99-111
in Student Module

(2)   You have lots of homework to do tonight. What choice best describes what you will probably do?

     (A)   You do what you want and forget to complete the homework.

     (B)   Start it, but quit when it gets hard.

     (C)   Wait until the last minute, but do it.

     (D)   Do it right away.

(3)   You are scheduled to work, but there is a party you would like to go to. What choice best describes what you will probably do?

     (A)   You go to the party and don't go to work.

     (B)   You call in sick and go to the party.

     (C)   You make an excuse to leave work an hour or two early so you can go to the party.

     (D)   You work your entire shift and only go to the party if it is still on after work.

(4)   Your parents ask you to do the dishes. What choice best describes what you will probably do?

     (A)   You ignore them and don't do the dishes.

     (B)   You say you'll get to them, but never "find the time."

     (C)   You procrastinate, but finally get them done.

     (D)   You do them quickly and without your parents nagging you.

(5)   You have a curfew of midnight. What choice best describes what you will probably do?

    (A)   You don't call and come home when it suits you.

    (B)   You finally call after you are already late and make excuses.

    (C)   You call to tell your parents you will be late.

    (D)   You get home on time.

(6)   The cashier mistakenly gives you an extra \$10 in change. What choice best describes what you will probably do?

    (A)   You keep the extra money and spend it.

    (B)   You keep it, but feel guilty about it.

    (C)   You start to keep it, but then decide to return it.

    (D)   You give it back as soon as you realize you have it.

(7)   You have a major paper due in one month. What choice best describes what you will probably do?

    (A)   You never do it.

    (B)   You finish it under pressure after the due date.

    (C)   You stay up late to finish it the night before it is due.

    (D)   You plan ahead and get it done on time without rushing.

A total of 60 students at the Milwaukee school responded to the survey and a total of 56 students from the school in Sweden responded. The sample at each school was selected in a way that was designed to produce a representative sample of the students at that school. The survey data are given in the tables that appear at the end of this investigation.

In this investigation, we will see how the data from these two samples can be used to decide if there is a "significant difference" in the proportion of students at the Milwaukee school and the proportion of students at the school in Sweden who respond in a particular way to one of these questions. By "significant difference" we mean that the difference in the response proportions for the two samples is larger than we would have expected to see just as a result of sampling variability.

The investigation will first guide you through the steps necessary to answer the following question:

> Is there a significant difference in the proportion of students who report that they would probably get home on time when given a curfew of midnight for students at the two schools?

After seeing how this question is answered, you will then be asked to answer an additional question using a similar process.

1. Let's consider the question posed: Is there a significant difference in the proportion of students who report that they would probably get home on time when given a curfew of midnight for students at the two schools?

To answer this question, we will need to look at the responses to question number 5 on the survey. In particular, we are interested in the proportions of students who answer D to this question. Answers A, B, and C can be combined into a single category here because they all represent answers where the student says he or she will probably not get home on time. For each of the two samples, we then need to know how many students answered D to question 5 and how many gave an answer other than D. We can use the data sets that appear at the end of this investigation to fill in the cells in the table below.

	Home on Time	Not Home on Time	Total
Milwaukee Sample			**60**
Sweden Sample			**56**
Total			**116**

Looking at the Milwaukee data set, we can count the number of D answers in the question 5 column. Verify that this number is 19 and enter it into the table above. Use the data sets provided to fill in the remaining cells and compare your numbers to those in the following table.

	Home on Time	Not Home on Time	Total
Milwaukee Sample	19	40	**60**
Sweden Sample	14	42	**56**
Total	**33**	**83**	**116**

2. We can now compute the relevant sample proportions. The proportion who report that they would probably be home on time for the Milwaukee sample is 19/60 = .32. Verify that the corresponding proportion for the Sweden sample is .25.

3. Notice that the two sample proportions are different: .32 for the Milwaukee sample and .25 for the Sweden sample. The difference between the two sample proportions is .07. How can we tell if this difference is significant? We know that even if there was no difference in the true proportions for the two schools, we would still not expect the sample proportions to be exactly equal because of sampling variability (those chance differences that occur from one sample to another as a result of the random selection process). To determine if a difference of .07 is significant, we need to know something about what kinds of differences are consistent with sampling variability alone.

As you might have guessed based on the previous two investigations, we will carry out a simulation. In this simulation, we create a population consisting of 116 individuals (representing the 116 survey respondents). Do this by cutting out 116 squares of paper that are the same size. You can use the templates at the end of this investigation if you don't want to make your own. Mark 33 of the 116 squares with an "H" to represent the 33 survey participants who reported they would probably be home on time. The other 83 slips of paper, which are unmarked, will represent the 83 survey participants who reported that they would probably not be home on time.

Place all the slips of paper in a paper bag and mix them well. Next, select 60 slips of paper from the bag to represent the 60 Milwaukee students. Count the number of the 60 selected squares that have an H and enter the count into the highlighted cell in the table below.

	Home on Time	Not Home on Time	**Total**
Milwaukee Sample			**60**
Sweden Sample			**56**
Total	**33**	**83**	**116**

Use the row and column totals to fill in the remaining cells in the table. Then, compute the two sample proportions and the difference between the two sample proportions:

Home on Time proportion of Milwaukee sample =

Home on Time proportion of Sweden sample =

Difference in sample proportions (Milwaukee – Sweden) =

172

For example, if the number in the highlighted cell was 16, the resulting table, proportions, and difference would be the following:

	Home on Time	Not Home on Time	Total
Milwaukee Sample	16	44	60
Sweden Sample	17	39	56
Total	33	83	116

Home on Time proportion of Milwaukee sample = 16/60 = .27

Home on Time proportion of Sweden sample = 17/56 = .30

Difference in sample proportions (Milwaukee − Sweden) = .27 − .30 = − .03

4. Record the difference you obtained in step 3 in the Trial 1 row of the table below.

Trial Number	Difference in Sample Proportions
1	
2	
3	
4	
5	

5. Put all of the slips of paper back in the bag, mix well, and then repeat step 3 four more times. Use the templates below to keep track of your results and to compute the relevant proportions. Finally, enter the resulting differences in sample proportions into the trials 2 through 5 rows of the table in step 4.

Trial 2			
	Home on Time	Not Home on Time	Total
Milwaukee Sample			60
Sweden Sample			56
Total	33	83	116

Home on Time proportion of Milwaukee sample =

Home on Time proportion of Sweden sample =

Difference in sample proportions (Milwaukee − Sweden) =

Trial 3

	Home on Time	Not Home on Time	Total
Milwaukee Sample			**60**
Sweden Sample			**56**
Total	**33**	**83**	**116**

Home on Time proportion of Milwaukee sample =

Home on Time proportion of Sweden sample =

Difference in sample proportions (Milwaukee – Sweden) =

Trial 4

	Home on Time	Not Home on Time	Total
Milwaukee Sample			**60**
Sweden Sample			**56**
Total	**33**	**83**	**116**

Home on Time proportion of Milwaukee sample =

Home on Time proportion of Sweden sample =

Difference in sample proportions (Milwaukee – Sweden) =

Trial 5

	Home on Time	Not Home on Time	Total
Milwaukee Sample			**60**
Sweden Sample			**56**
Total	**33**	**83**	**116**

Home on Time proportion of Milwaukee sample =

Home on Time proportion of Sweden sample =

Difference in sample proportions (Milwaukee – Sweden) =

6. OK, we are almost there! Each student in your class should have carried out five trials. Now combine all the simulated differences into one large data set by recording the values obtained by each student in your class in the table on the next page.

7. Use the data from step 6 to construct a dotplot of the simulated differences.

---

8. The dotplot of the simulated differences in step 7 provides information about how large the difference in sample proportions can be expected to be just as a result of sampling variability. That is, these simulated sample differences are consistent with the situation where there is no difference in the population proportions.

Now, let's go back and look at the actual difference in sample proportions for the Milwaukee and Sweden samples. The observed difference was .07. Based on what you see in the dotplot of simulated differences, is sampling variability a plausible explanation for why we might see a difference in sample proportions as large as .07? Explain why or why not.

9. Write a few sentences that address the original question posed: Is there a significant difference in the proportion of students who report that they would probably get home on time when given a curfew of midnight for students at the two schools?

10. In the last part of this investigation, your teacher will assign you to a team. Each team will either choose one of the questions below or formulate a question of its own that can be answered using the Milwaukee and Sweden data. Once your teacher has approved the team's question, use the data provided to answer it and write a brief report summarizing your results. (You should use a process similar to the one used to answer the "home on time" question, but note that because only the members of your team will be carrying out the simulation, you will each need to carry out more than five trials.)

Some possible questions to investigate:

Is there a significant difference in the proportion of students who report that they would give back an extra $10 in change for students at the two schools?

Is there a significant difference in the proportion of students who report that they would plan ahead and get a paper done without rushing and on time for students at the two schools?

Is there a significant difference in the proportion of students who report that they do homework right away for students at the two schools?

Data Set: Milwaukee

Student #	Question (1)	Question (2)	Question (3)	Question (4)	Question (5)	Question (6)	Question (7)
1	Female	A	B	C	C	A	B
2	Female	D	D	D	D	B	D
3	Female	C	D	D	C	A	C
4	Female	A	D	D	D	C	C
5	Female	C	D	D	C	A	D
6	Female	C	D	D	D	D	C
7	Female	C	D	C	C	B	C
8	Female	D	D	C	C	C	C
9	Female	B	C	C	C	A	C
10	Female	D	C	C	B	A	C
11	Female	C	D	C	C	C	D
12	Female	C	B	C	C	D	C
13	Female	D	D	D	D	D	D
14	Male	A	D	C	C	C	B
15	Female	D	D	D	D	D	C
16	Male	D	B	C	D	D	C
17	Male	C	D	C	C	D	C
18	Male	D	D	D	D	D	D
19	Male	C	D	C	B	C	C
20	Male	C	B	D	C	A	C
21	Female	A	C	B	C	A	C
22	Male	C	D	D	B	D	C
23	Male	C	D	D	B	D	C
24	Male	C	D	A	A	B	C
25	Male	C	D	C	D	A	D
26	Female	C	D	D	C	D	C
27	Female	C	D	C	C	C	C
28	Male	C	D	A	C	D	C
29	Male	A	D	B	C	B	C
30	Female	C	C	C	B	C	C

Data Set: Milwaukee							
Student #	Question (1)	Question (2)	Question (3)	Question (4)	Question (5)	Question (6)	Question (7)
31	Male	D	D	D	C	C	D
32	Male	B	B	C	B	D	D
33	Female	C	C	B	A	A	D
34	Female	C	D	C	C	A	C
35	Male	C	D	D	C	D	C
36	Male	B	C	C	D	A	C
37	Male	C	D	D	A	D	C
38	Female	D	D	D	D	D	D
39	Female	C	C	D	C	A	C
40	Male	C	D	B	C	D	C
41	Male	C	D	A	C	A	D
42	Male	C	B	C	C	D	C
43	Female	C	D	D	D	D	C
44	Male	D	D	C	C	B	D
45	Male	A	A	A	A	A	C
46	Female	C	D	C	D	D	C
47	Female	C	D	C	D	C	C
48	Male	D	C	A	D	D	C
49	Male	A	B	B	B	A	C
50	Female	C	D	B	B	A	C
51	Female	D	D	B	D	B	C
52	Male	B	C	B	A	A	C
53	Male	A	D	B	D	A	C
54	Female	D	D	C	C	C	D
55	Male	C	D	C	D	A	D
56	Female	A	D	D	D	C	C
57	Male	C	D	C	C	A	C
58	Female	C	C	B	C	C	C
59	Male	C	D	D	C	B	D
60	Female	C	D	D	D	D	C

Data Set: Sweden

Student #	Question (1)	Question (2)	Question (3)	Question (4)	Question (5)	Question (6)	Question (7)
1	Female	C	D	C	B	D	C
2	Female	C	D	C	B	D	C
3	Female	B	D	B	B	D	C
4	Male	C	C	B	B	A	C
5	Male	C	B	D	D	A	C
6	Female	C	C	B	C	A	C
7	Female	C	B	B	B	B	C
8	Male	A	B	D	C		C
9	Female	D	D	D	C	C	D
10	Female	C	C	D	C	B	C
11	Female	C	D	D	D	D	C
12	Male	B	D	D	D	B	C
13	Female	C	D	B	D	D	D
14	Female	D	C	B	C	D	D
15	Female	D	C	C	C	A	D
16	Male	D	C	D	C	C	C
17	Male	C	D	D	D	A	D
18	Male	C	C	C	C	A	C
19	Female	C	D	C	C	D	C
20	Female	D	C	D	C	A	D
21	Female	D	D	C	D	A	C
22	Male	C	B	C	C	A	C
23	Female	C	D	D	C	D	C
24	Female	B	C	C	B	A	C
25	Male	D	C	C	D	A	C
26	Female	C	D	D	C	D	D
27	Female	C	D	D	C	A	C
28	Female	D	D	C	D	D	C

<table>
<tr><td colspan="8" align="center">Data Set: Sweden</td></tr>
<tr><th>Student #</th><th>Question (1)</th><th>Question (2)</th><th>Question (3)</th><th>Question (4)</th><th>Question (5)</th><th>Question (6)</th><th>Question (7)</th></tr>
<tr><td>29</td><td>Female</td><td>D</td><td>D</td><td>D</td><td>D</td><td>C</td><td>C</td></tr>
<tr><td>30</td><td>Female</td><td>D</td><td>D</td><td>C</td><td>C</td><td>C</td><td>C</td></tr>
<tr><td>31</td><td>Male</td><td>B</td><td>D</td><td>C</td><td>C</td><td>D</td><td>D</td></tr>
<tr><td>32</td><td>Female</td><td>C</td><td>D</td><td>B</td><td>D</td><td>A</td><td>C</td></tr>
<tr><td>33</td><td>Female</td><td>C</td><td>D</td><td>B</td><td>C</td><td></td><td>C</td></tr>
<tr><td>34</td><td>Male</td><td>C</td><td>D</td><td>D</td><td>C</td><td>A</td><td>B</td></tr>
<tr><td>35</td><td>Female</td><td>C</td><td>D</td><td>D</td><td>C</td><td>D</td><td>C</td></tr>
<tr><td>36</td><td>Female</td><td>B</td><td>D</td><td>D</td><td>C</td><td>B</td><td>D</td></tr>
<tr><td>37</td><td>Female</td><td>C</td><td>D</td><td>D</td><td>C</td><td>B</td><td>C</td></tr>
<tr><td>38</td><td>Male</td><td>C</td><td>C</td><td>C</td><td>B</td><td>D</td><td>C</td></tr>
<tr><td>39</td><td>Female</td><td>C</td><td>C</td><td>B</td><td>C</td><td>A</td><td>C</td></tr>
<tr><td>40</td><td>Female</td><td>C</td><td>C</td><td></td><td>C</td><td>D</td><td>C</td></tr>
<tr><td>41</td><td>Female</td><td>C</td><td>C</td><td>D</td><td>C</td><td>B</td><td>D</td></tr>
<tr><td>42</td><td>Female</td><td>B</td><td>C</td><td>D</td><td>C</td><td>A</td><td>B</td></tr>
<tr><td>43</td><td>Male</td><td>D</td><td>C</td><td>C</td><td>C</td><td>A</td><td>D</td></tr>
<tr><td>44</td><td>Female</td><td>C</td><td>D</td><td>C</td><td>D</td><td>B</td><td>D</td></tr>
<tr><td>45</td><td>Female</td><td>C</td><td>C</td><td>C</td><td>C</td><td>B</td><td>C</td></tr>
<tr><td>46</td><td>Male</td><td>C</td><td>D</td><td>D</td><td>C</td><td>A</td><td>C</td></tr>
<tr><td>47</td><td>Male</td><td>C</td><td>D</td><td>D</td><td>C</td><td>A</td><td>C</td></tr>
<tr><td>48</td><td>Female</td><td>C</td><td>C</td><td>D</td><td>D</td><td>A</td><td>D</td></tr>
<tr><td>49</td><td>Male</td><td>C</td><td>C</td><td>C</td><td>B</td><td>A</td><td>C</td></tr>
<tr><td>50</td><td>Female</td><td>D</td><td>D</td><td>C</td><td>C</td><td>D</td><td>D</td></tr>
<tr><td>51</td><td>Male</td><td>C</td><td>D</td><td>D</td><td>D</td><td>D</td><td>D</td></tr>
<tr><td>52</td><td>Male</td><td>C</td><td>D</td><td>D</td><td>D</td><td>D</td><td>C</td></tr>
<tr><td>53</td><td>Male</td><td>B</td><td>C</td><td>D</td><td>C</td><td>A</td><td>C</td></tr>
<tr><td>54</td><td>Male</td><td>C</td><td>D</td><td>D</td><td>C</td><td>C</td><td>C</td></tr>
<tr><td>55</td><td>Female</td><td>D</td><td>D</td><td>D</td><td>B</td><td>B</td><td>D</td></tr>
<tr><td>56</td><td>Female</td><td>C</td><td>C</td><td>D</td><td>C</td><td>D</td><td>C</td></tr>
</table>

Random Number Table

```
7 8 7 2 1    7 2 6 4 7    0 4 8 9 9    6 5 9 4 8
1 6 5 4 2    0 8 6 3 3    7 5 6 2 6    8 4 0 0 6
6 9 7 0 9    6 8 2 1 2    6 5 7 7 8    8 0 2 8 1
5 6 9 7 7    0 8 7 2 7    6 4 5 3 9    5 8 9 9 7
7 8 7 5 1    9 0 2 4 6    4 9 2 5 5    1 9 4 0 8
4 3 3 1 0    4 4 8 9 4    4 3 4 3 5    5 5 7 9 3
1 6 7 8 0    4 5 8 1 4    8 9 6 0 3    0 8 3 2 8
9 8 5 9 9    8 1 2 8 3    2 3 3 9 9    0 5 2 7 7
1 0 1 8 2    0 0 2 3 1    9 3 4 6 7    8 0 1 4 0
6 8 8 2 2    5 0 7 7 4    1 6 7 8 1    6 7 9 2 3
6 3 1 6 3    8 2 5 1 1    5 7 3 2 6    0 7 2 0 6
4 7 5 2 1    2 4 4 7 3    1 1 1 7 0    8 0 0 0 8
9 8 0 3 0    9 2 7 0 6    3 0 5 2 2    4 3 4 0 6
3 8 3 2 5    5 4 1 6 1    0 9 1 7 6    3 3 7 9 0
3 6 4 0 6    8 0 5 0 2    0 1 4 1 0    2 1 1 5 1
5 7 0 7 1    7 5 5 3 0    0 9 2 3 8    8 9 1 1 8
3 2 8 4 4    2 6 7 5 6    8 6 4 8 6    7 6 1 0 7
0 8 9 9 7    2 4 9 7 6    8 5 4 3 8    8 3 9 4 1
1 9 4 2 0    4 3 6 5 0    5 4 2 8 0    7 3 1 0 7
3 6 7 7 7    8 9 4 2 8    6 3 5 3 2    9 9 0 7 9
3 0 9 9 0    7 8 9 8 3    3 0 9 7 0    9 3 3 0 5
7 3 8 9 5    5 9 4 9 3    3 3 3 1 3    7 1 8 6 0
5 3 5 6 1    7 5 6 4 9    1 2 4 4 8    4 9 9 2 2
7 5 2 6 4    1 4 4 3 2    3 9 2 8 8    7 3 7 8 9
9 8 1 6 7    0 2 9 6 0    4 2 4 3 2    9 5 1 7 1
0 8 0 3 8    6 4 5 3 1    1 4 5 0 1    5 6 0 1 5
7 8 1 6 0    3 3 0 5 3    6 8 2 1 8    8 5 4 2 7
5 5 9 9 7    1 0 4 6 9    3 4 5 9 5    9 3 3 3 4
1 6 5 8 1    5 6 4 8 6    7 4 4 4 4    8 1 7 7 5
2 7 9 4 5    6 2 7 9 5    8 9 4 4 8    9 9 5 5 7
8 0 6 0 1    7 6 0 2 8    9 9 0 1 8    3 1 6 6 6
5 7 6 0 4    6 9 4 5 5    7 0 5 3 7    9 6 8 4 2
4 5 3 2 5    2 0 8 5 5    0 5 1 0 6    4 2 5 6 2
3 7 9 3 2    0 6 6 6 6    6 5 7 1 5    7 8 7 1 4
0 3 1 3 5    7 8 0 6 8    7 2 4 8 3    1 1 0 9 9
7 4 9 7 2    4 7 1 7 5    2 3 3 1 4    4 6 7 5 2
2 8 3 7 3    9 0 0 1 1    7 8 1 9 5    0 0 1 3 9
5 2 8 3 1    3 0 2 5 9    0 6 6 2 1    2 7 7 2 5
9 5 1 3 8    0 0 3 3 8    7 1 1 7 0    6 8 6 0 4
5 4 3 5 0    3 8 2 2 7    0 6 8 0 5    1 9 7 6 4
0 1 2 9 7    5 1 4 3 9    5 8 8 6 7    7 4 3 7 6
2 6 1 2 0    0 8 1 2 1    7 7 5 4 0    4 8 0 1 6
4 0 6 8 9    9 6 9 0 3    6 6 5 2 6    2 2 4 2 0
```

Made in the USA
Monee, IL
07 July 2026

56549107R00109